# Carbon-13 NMR in Polymer Science

# Carbon-13 NMR in Polymer Science

**Wallace M. Pasika,** EDITOR

*Laurentian University*

Based on a symposium

sponsored by the

Macromolecular Science

Division at the 61st

Conference of The Chemical

Institute of Canada in

Winnipeg, Manitoba,

June 4–7, 1979.

ACS SYMPOSIUM SERIES **103**

AMERICAN CHEMICAL SOCIETY

WASHINGTON, D. C.       1979

Library of Congress CIP Data

Carbon-13 NMR in polymer science.
  (ASC symposium series; 103 ISSN 0097–6156)

  Includes bibliographies and index.

  1. Nuclear magnetic resonance spectroscopy—Congresses. 2. Carbon—Isotopes—Spectra—Congresses. 3. Polymers and polymerization—Spectra—Congresses.
  I. Pasika, Wallace M. II. Chemical Institute of Canada. Macromolecular Science Division. III. Series: American Chemical Society. ACS symposium series; 103.

QD272.S6C37        547'.84            79-13384
ISBN 0–8412–0505–1      ACSMC8 103 1–344 1979

PRINTED IN THE UNITED STATES OF AMERICA

# ACS Symposium Series

## Robert F. Gould, *Editor*

# FOREWORD

The ACS SYMPOSIUM SERIES was founded in 1974 to provide a medium for publishing symposia quickly in book form. The format of the Series parallels that of the continuing ADVANCES IN CHEMISTRY SERIES except that in order to save time the papers are not typeset but are reproduced as they are submitted by the authors in camera-ready form. Papers are reviewed under the supervision of the Editors with the assistance of the Series Advisory Board and are selected to maintain the integrity of the symposia; however, verbatim reproductions of previously published papers are not accepted. Both reviews and reports of research are acceptable since symposia may embrace both types of presentation.

# CONTENTS

# PREFACE

Carbon-13 NMR is opening up new vistas in polymer chemistry. Analytical information on polymeric systems, information regarding their structural, dynamic, and polymerization characteristics, can be provided by C-13 NMR. In many instances C-13 NMR complements already familiar characterization techniques or allows characterization to be carried out more readily. In other instances, only C-13 NMR can provide answers.

While organizing the symposium upon which this volume is based, the Macromolecular Science Division of the Chemical Institute of Canada attempted to include papers representing a wide range of applications for using Carbon-13 NMR to characterize polymers and to have both synthetic and biomacromolecular systems considered.

Many contributed to the success of the symposium. The executive and members of the Macromolecular Science Division thank the following firms for fiscal support: Polysar Limited; Abitibi Paper Company; Reichhold Limited; Xerox Research Centre of Canada Limited; Varian Associates of Canada; Glidden Company; Domtar Limited; DuPont of Canada; Shell Canada; Gulf Oil Canada; and the Dunlop Research Centre. A special vote of thanks is extended to the speakers at the symposium and to the authors for their excellent presentations and for their cooperation in "putting it together." J. Comstock of the American Chemical Society showed understanding and patience in bringing this volume to print.

Laurentian University
Sudbury, Ontario
Canada
March 26, 1979

WALLACE M. PASIKA

# The Study of the Structure and Chain Dynamics of Polysulfones by Carbon-13 NMR

F. A. BOVEY and R. E. CAIS

Bell Laboratories, Murray Hill, NJ 07974

Sulfur dioxide does not homopolymerize but does participate in a rather wide variety of free radical copolymerizations with unsaturated monomers. The resulting polysulfones have been known for quite a long time. Solonina ($\underline{1},\underline{2}$) obtained a white solid from the reaction of $SO_2$ with allyl ethers in 1898, but such products were not recognized as copolymers until the work of Marvel and Staudinger in the 1930's.

The class of vinyl monomers which will copolymerize with $SO_2$ is not clearly distinguishable in terms of fundamental structural characteristics from the class of those that will not. All terminal olefins, beginning with ethylene and continuing on up to the higher olefins, copolymerize and, with the exception of ethylene, give copolymers of strictly 1:1 alternating structure, since neither the chains ending in $SO_2$ nor those ending in olefin can add their own monomer; i.e. both reactivity ratios $r_1$ and $r_2$ are zero. Others include cyclopentene, cyclohexene, cycloheptene and some open-chain olefins with internal double bonds, provided the substitutents at the double bond are not all larger than methyl.

Monomers which can add to their own radicals are capable of copolymerizing with $SO_2$ to give products of variable composition. These include styrene and ring-substituted styrenes (but not α-methylstyrene), vinyl acetate, vinyl bromide, vinyl chloride, and vinyl floride, acrylamide (but not N-substituted acrylamides) and allyl esters. Methyl methacrylate, acrylic acid, acrylates, and acrylonitrile do not copolymerize and in fact can be homopolymerized in $SO_2$ as solvent. Dienes such as butadiene and 2-chlorobutadiene do copolymerize, and we will be concerned with the latter compound in this discussion.

An important feature of olefin-$SO_2$ polymerizations is their relatively low <u>ceiling temperatures</u>: the reverse reaction becomes evident at quite low temperatures. In fact, it is while studying such copolymers that Dainton and Ivin in 1948 ($\underline{3}$) first clearly recognized the existence of ceiling temperatures in vinyl polymers. The reversibility of the propagation has important effects on the composition of the chains, which, as we shall see, exhibits a very

0-8412-0505-1/79/47-103-001$06.25/0

strong dependence on temperature even at constant monomer feed
ratio.  This dependence invariably takes the form of tending to
exclude $SO_2$ as the temperature is increased.  This reversibility
accounts also for some other unusual features of the chain structure, including deviation from first-order Markov statistics and,
in the case of chloroprene, a seemingly anomalous and rather
interesting tendency to form a more regular chain structure as the
polymerization temperature is increased.

The use of nmr, --carbon-13 nmr in particular--, has given a
much deeper insight into the structure of $SO_2$ copolymers than was
possible by the older, traditional method of analytically determining monomer ratios in the polymer as a function of the monomer
feed.  In fact, it can be safely said that the use of nmr has
completely revolutionized the study of copolymers.  (The impact of
nmr on copolymer studies is studiously ignored in all polymer
textbooks, which tend to reflect the status of the field twenty
years ago.)

Styrene-$SO_2$ Copolymers.  I would now like to discuss two systems which illustrate the power of C-13 nmr in structural studies.
The first is the styrene-$SO_2$ system.  As already indicated, this
is of the type in which the chain composition varies with monomer
feed ratio and also with temperature at a constant feed ratio (and
probably with pressure as well.)  The deviation of the system from
simple, first-order Markov statistics, --i.e. the Lewis_Mayo
copolymerization equation--, was first noted by Barb in 1952 (4)
who proposed that the mechanism involved complex formation between
the monomers.  This proposal was reiterated about a decade later
by Matsuda and his coworkers (5,6).  Such charge transfer complexes do in fact exist, with the olefin acting as the donor, but
we shall see that it is not necessary to invoke complex formation
to explain the observed kinetics.  It is also unnecessary to
invoke penultimate effects in the reaction of the propagating
radicals.

The traditional compositional and kinetic measurements cannot
distinguish effectively between the various models proposed to
account for deviations from the simple copolymerization model.  To
do this requires monomer sequence data, and for this nmr is the
method par excellence.  But in order to make use of the potential-
ly rich information provided by nmr, one must be able to make
assignments of the resonances in spectra which are often quite
complex.  This is usually done by (1) isotopic labelling; (2)
observing carbon multiplicity in the absence of decoupling; (3)
relaxation measurements; (4) logical deduction from the spectra
of a series of copolymers of varied composition; and (5) invoking
chemical shift rules.  In order to interpret the spectra of $SO_2$
copolymers, we must recognize certain features of the monomer
sequences, the principal one of which is that they have a sense of
direction.  To avoid confusion, we must remember that for any
particular sequence we are always looking at the central vinyl
monomer unit in the β-to-α direction from left to right.  A

second, minor point is that since $SO_2$ units are not chiral, compositional triads are actually configurational dyads and compositional tetrads are configurational triads. These points are illustrated in Figures 1 and 2, where the chain sequences are represented as planar zigzags (viewed edge on). Figure 1 shows compositional triads involving the unit X, which in our case is $SO_2$. As configurational dyads they may be <u>racemic</u> or <u>meso</u> of two types, depending on whether the chiral centers straddle an X unit or not. Also, we must note that XMM is not the same as MMX. In Fig. 2 the same representation is extended to compositional tetrads, i.e. configurational triads.

An important first step in interpreting the C-13 spectra is to distinguish α-carbons from β-carbons, i.e. methine from methylene. Observation of multiplicity when the proton decoupler is off is one way, but this is not always easy if the lines are broadened by chemical shift multiplicity. Measurement of $T_1$ has been used for this purpose since the β-carbon with two bonded protons relaxes about twice as fast as the α-carbon with only one. A very positive way is by deuterium labelling. In Fig. 3 is shown the main-chain 25 MHz carbon spectrum of two styrene-$SO_2$ copolymers containing 58 mol% styrene, or a ratio of styrene to $SO_2$ of 1.38 (<u>7</u>). In the bottom one, β,β-$d_2$-styrene has been used, and all the β-carbon resonances are distinguishable from the α-carbon resonances since the presence of deuterium has eliminated their nuclear Overhauser effect; because of this and the deuterium J coupling (∿20 Hz), they are markedly smaller and broader than the α-carbon resonances.

The assignments of resonances to particular monomer sequences are based primarily on their relative intensities as a function of overall composition. We define R as the overall ratio of styrene to $SO_2$ in the polymer. In Fig. 4 are shown the complete 25 MHz C-13 spectra, including at the left the aromatic carbons, for four copolymers of varied R (<u>7</u>). The spectrum of atactic polystyrene is also shown. The number of resonances shows that compositional <u>triads</u> are being distinguished: SMS, SMM, MMS, and MMM. Here, M stands for styrene and S for $SO_2$. As the ratio R increases, MMM sequences become evident and can be assigned on the basis of the polystyrene spectrum. In the latter, the β-carbon is highly sensitive to configuration whereas the α-carbon is entirely insensitive, appearing as a narrow spike.

In Table 1 are shown the chemical shifts of the resonances in polystyrene sulfones. They are based on relative resonance intensities as a function of monomer ratio and on a consistent set of rules. We observe that the least shielded α- and β-carbons are those bonded directly to sulfone sulfur atoms. One can recognize very clearly a sulfone oxygen shielding effect when the carbon concerned is in the γ-position with relation to the oxygens. It appears to be analagous to the well known carbon γ-effect. Comparing α-MMS with α-SMS we see that for α-carbons it causes a shielding of about 5 ppm. Comparing β-SMM with β-SMS, we see a

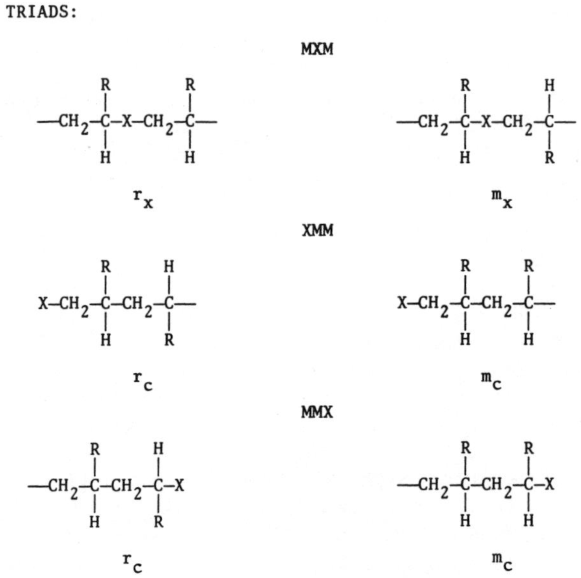

TRIADS:

Figure 1.   Compositional triads (configurational dyads) in chains of vinyl (M) copolymers with a comonomer which places a single atom X in the main chain

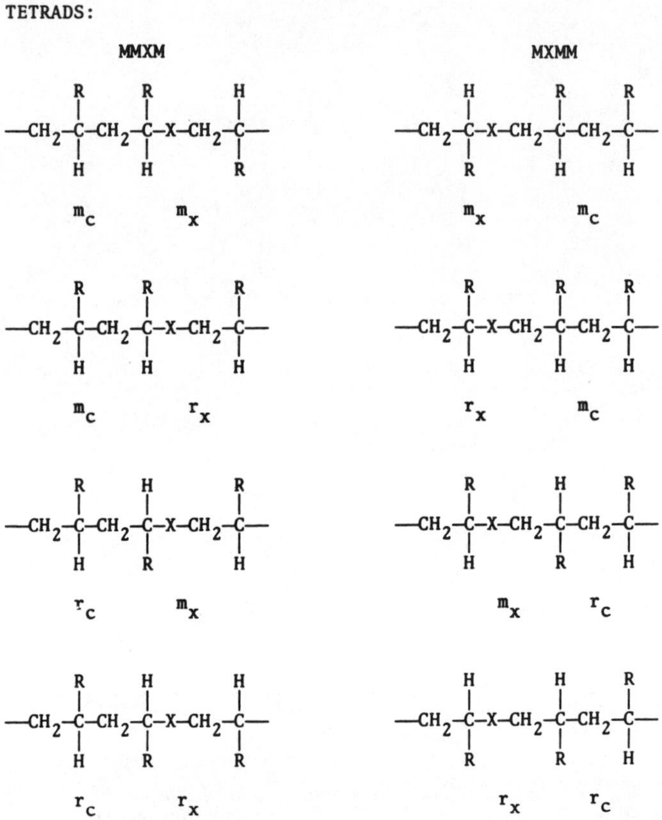

TETRADS:

Figure 2.  Compositional tetrads (configurational triads) in chains of vinyl (M) copolymers with a comonomer which places a single atom X in the main chain.

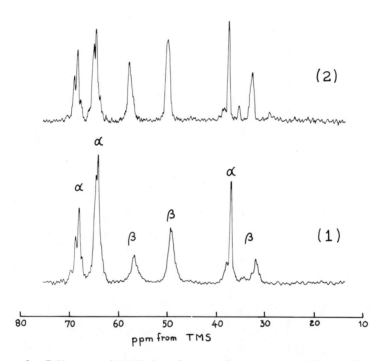

Figure 3.   C-13 spectra (25 MHz) of the main chain of styrene–SO₂ copolymers containing 58 mol % styrene (R = 1.38).  The bottom spectrum is of a copolymer with β,β-d₂-styrene  (25% solution in CDCl₃ at 55°C).

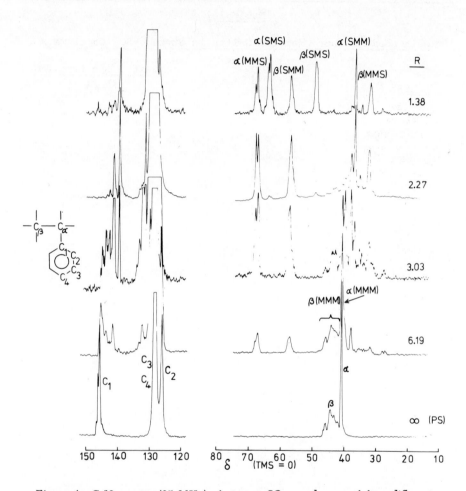

*Figure 4.   C-13 spectra (25 MHz) of styrene–SO₂ copolymers of four different compositions. The aromatic carbons are on the left and the main chain carbons on the right. R values are the mole ratios of styrene to SO₂ in the copolymers. The spectrum at the bottom is that of atactic PS. (All observed as 25% solutions in CDCl₃ at 55°C except PS, which was observed as 20% solution in cyclohexane-d₁₂ at 77°C.)*

shielding of about 9 ppm for β-carbons.  The same is true for β-MMM vs. β-MMS.  We note that α-SMM and β-MMS are actually more shielded than polystyrene itself.

---

## Table 1

### Main Chain Carbon Chemical Shifts in Styrene-SO₂ Copolymers

| | | Shift, ppm from TMS |
|---|---|---|
| α-MMS | 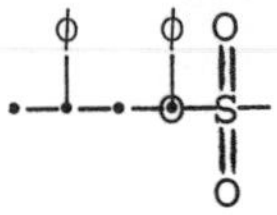 | 68 |
| α-SMS | 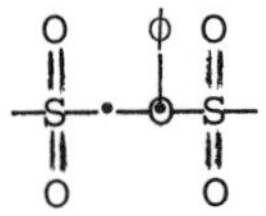 | 63 |
| β-SMM | 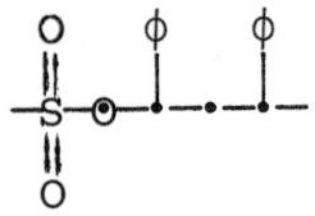 | 57 |
| β-SMS | 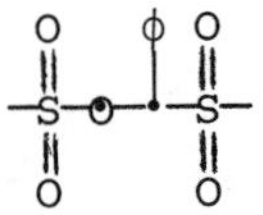 | 48 |
| β-MMM | 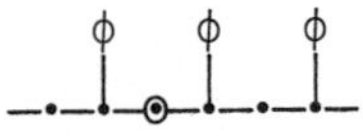 | 44 |
| α-MMM | 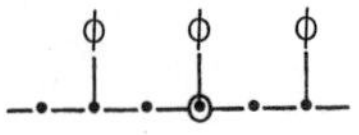 | 40 |
| α-SMM | 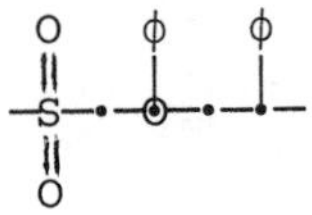 | 38 |
| β-MMS | 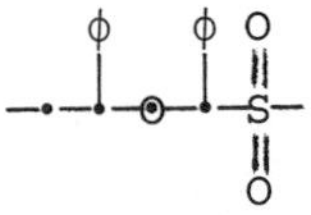 | 33 |

---

An interesting feature of the styrene-SO₂ system, --which indeed is true of all SO₂ copolymerizations with comonomers capable of homopolymerizing--, is the existence of a ceiling temperature above which the formation of alternating units, SMS, is forbidden.  The number fraction of M sequences of length n is given by;

$$N_M(n) = \frac{p(SM^nS)}{\sum\limits_{n=1}^{\infty} p(SM^nS)}$$

i.e. the probability of occurrence of a length of n styrene units divided by the sum of the probabilities of all styrene sequences of all lengths.  Since it can be shown that:

$$\sum\limits_{n=1}^{\infty} p(SM^nS) = p(MS),$$

and since in the present case:

$$p(MS) = p(S),$$

because all sulfone units S have M as neighbors, we have

$$N_M(n) = \frac{p(SM^nS)}{p(S)}$$

For the number fraction of SMS sequences:

$$N_M(1) = \frac{p(SMS)}{p(S)}$$

A plot of $N_M(1)$ versus polymerization temperature is shown in Fig. 5.  It will be seen that at low temperature it approaches unity, which would correspond to a strictly alternating structure, and that it declines very rapidly, reaching zero at about 40°. Therefore, above 40° alternating sequences are not generated.  At 40°, it turns out that a chain ending in -SM· will add another M and then add S, and so on, yielding a polymer with predominantly the regular structure SMMSMMSMM etc.  As the polymerization temperature is increased further, the $SO_2$ units are in effect squeezed out progressively until around 100° a second ceiling temperature occurs and no appreciable amount of $SO_2$ is incorporated; only polystyrene is produced.  There is probably a moral here with regard to those monomers which, like methyl methacrylate, acrylonitrile, and acrylate esters, merely seem to homopolymerize in $SO_2$ as solvent.  If polymerization were conducted at a sufficiently low temperature, $SO_2$ probably would be incorporated.
    The copolymerization scheme we favor is as follows:

$$\text{∿SM· + S} \underset{}{\overset{K_1}{\rightleftarrows}} \text{∿SMS·} \qquad\qquad (1)$$

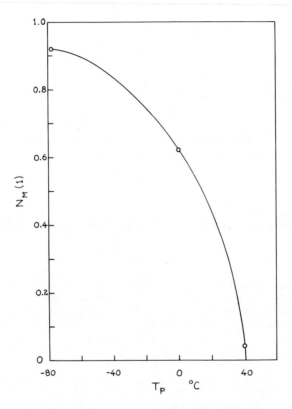

*Figure 5.   Plot of the number fraction of SMS sequences, $N_M(1)$ as a function of polymerization temperature for styrene–$SO_2$ copolymers*

$$\sim MS\cdot \ + \ M \ \underset{\leftarrow}{\overset{K_2}{\rightarrow}} \ \sim SM\cdot \qquad\qquad (2)$$

$$\sim SM\cdot \ + \ M \ \underset{\leftarrow}{\overset{k_p}{\rightarrow}} \ \sim MM\cdot \qquad\qquad (3)$$

$$\sim MM\cdot \ + \ M \ \underset{\leftarrow}{\overset{k_p{}'}{\rightarrow}} \ \sim MM\cdot \qquad\qquad (4)$$

$$\sim MM\cdot \ + \ S \ \underset{\leftarrow}{\overset{K_3}{\rightarrow}} \ \sim MMS\cdot \qquad\qquad (5)$$

The K values are the equilibrium constants which describe the position of the propagation-depropagation equilibrium and are equal to the ratio of the propagation rate constant to the depropagation rate constant:

$$K = k_{propag.}/k_{depropag.}$$

$$K_1 = \frac{[\sim SMS\cdot]}{[\sim SM\cdot][S]}$$

$$K_2 = \frac{[\sim SM\cdot]}{[\sim SMS\cdot][M]}$$

The formation of alternating copolymer is represented by the overall equilibrium constant

$$K_{alt} = K_1 K_2 = \frac{1}{[S][M]}$$

As $K_{alt}$ approaches unity we approach the ceiling temperature for alternating copolymerization. At very low temperature, --for example -78°--, the alternating structure predominates because only S is reactive enough to add to $\sim M\cdot$ and $\sim S\cdot$, of course, can only add M. Above 40°, alternating sequences cannot survive.

Chloroprene-SO$_2$ Copolymers. The second copolymer system to be discussed is the chloroprene-SO$_2$ system. This presents potential complications beyond that of the styrene-SO$_2$ system because the chloroprene may enter the chain in 1,4-cis, 1,4-trans, 1,2 or 3,4 fashion:

1,4-cis                                        1,4-trans

1,2                                              3,4

If it enters in 1,2 or 3,4 fashion, chiral centers are introduced and relative stereoisomerism must be considered.  Actually,
as it turns out, things are not so complicated as they might be
because of a strong preference for one mode of addition.  Complications are quite sufficient, however, as they include a head-to-
head:head-to-tail option not normally open to vinyl monomers but
available to unsymmetrically substituted dienes adding in 1,4-
fashion.  In fact, little progress can be made in structural
determination except by use of C-13 nmr at superconducting frequencies.

We have prepared a series of copolymers under the conditions
shown in Table 2.  The monomer feed was always a 50:50 ratio of
chloroprene to sulfur dioxide.  Copolymerizations were carried out
in bulk at temperatures from -78° to 100°.  Initiators were tertiary butyl hydroperoxide at low temperatures, where it forms a
redox system with the $SO_2$ and is more effective than one might
otherwise expect.  Silver nitrate was used at 0° and 25°, azoiso-
butyronitrile at 40° and 60°, and azodicyclohexanecarbonitrile

at 100°.
R values were calculated from elemental analysis for carbon,
hydrogen, and chlorine.  It can be seen again that temperature
has a very marked effect on composition.  Even at 100°, however,
about 16  mol% sulfur dioxide is present.  There was also produced
a small quantity (1 to 10% of the amount of copolymer) of the
cyclic addition product, 3-chloro-2,5-dihydrothiophene-1,1-dioxide,
m.p. 99-100°.

We show in Scheme I the representation of the four chloroprene-
centered sequences MMS, MMM, SMS, SMM (M as before represents
monomer and S represents sulfur dioxide).  The chloroprene carbons
are numbered in the conventional manner.  The first two carbons of
chloroprene units which are neighbors of the central chloroprene
unit or of the sequence are also represented.  We will be concerned with all four carbons in all four sequences.

The spectrum of the polymer prepared at 40°, with an R value
of 1.64, is shown in Fig. 6.  The spectrum divides naturally into
two parts: that corresponding to the olefinic carbons, $C_2$ and $C_3$,

## Table 2

Preparation of Chloroprene-$SO_2$ Copolymers; All in Bulk at 50:50 Mol Ratio

| Temp. (°C) | Polym. Time (h) | Initiator | [Init.] (mol%) | Yield (wt%) | R | Notes |
|---|---|---|---|---|---|---|
| -78 | 17.25 | t-BuOOH | 0.46 | 9.8 | 1.15 | sol. DMSO, DMF |
| -45 | 1.67 | " | " | 14.0 | 1.20 | " |
| -17 | 6.25 | " | " | 15.4 | 1.20 | 2% sol. hot $CHCl_3$ |
| 0 | 14.80 | $AgNO_3$ | 0.02 | 22.6 | 1.20 | sol. DMSO, DMF |
| 25 | 4.67 | " | " | 22.3 | 1.30 | " |
| 40 | 6.32 | AIBN | 0.05 | 22.1 | 1.64 | " |
| 60 | 3.30 | " | " | 32.1 | 2.44 | 49% gel, 0.9% CDTD[2] |
| 100 | 1.03 | ACHCN[1] | " | 15.4 | 5.08 | 52% gel, 10.1% CDTD |

[1] 1,1-Azodicyclohexanecarbonitrile.

[2] 3-Chloro-2,5-dihydro-thiophene-1,1-dioxide.

Scheme I.   $X = Cl$ or $H$; $Y = Cl$ or $H$.

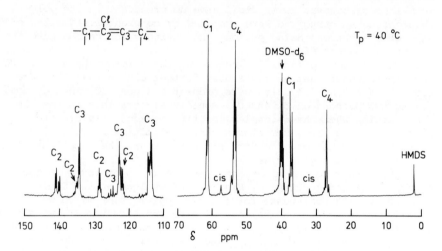

*Figure 6. A 90-MHz C-13 spectrum of chloroprene–SO₂ copolymer prepared at 40°C and having an R value of 1.64*

appears at the left between 110 and 150 ppm downfield from the
hexamethyldisiloxane reference, and the part corresponding to the
methylene carbons, $C_1$ and $C_4$, appears at the right in the more
shielded region of 25-65 ppm. We shall first assign the carbons,
then the compositional sequences, and finally concern ourselves
with head-to-head:head-to-tail isomerism.

The olefinic carbons $C_2$ and $C_3$ were assigned by (a) a partially relaxed dynamic experiment in which by proper choice of
pulse interval the $C_2$ carbon resonances, having longer $T_1$, re-
mained as inverted signals while the $C_3$ carbons recovered to
positive values; and (b) from the proton-coupled spectrum, in
which the $C_3$ resonances were split to doublets.

In the methylene region the $C_1$ and $C_4$ resonances at higher
field occur in the same position as for polychloroprene and the
assignements were taken from the work of Coleman et al. (8). The
methylene carbons at lower field have no precedent in the polychloroprene spectrum and were assigned by coherent irradiation of
each of the corresponding protons and noting which carbon gave the
larger Overhauser effect.

We must now consider the nature of the chloroprene units. We
have represented them as 1,4 and we further deduce that they are
trans. If 1,2 or 3,4 units were present the pendant vinyl groups,
having two protons on $C_1$ and $C_4$, would yield triplets in the olefinic region when the decoupler is off. No such triplets are
observed, as so these structures can be eliminated. This also
means the elimination of asymmetric carbons and the complications
to which they can give rise.

The answer to the cis-trans question is to be found in the
methylene carbon spectrum of Fig. 7. If we look at the $C_1$ (61
ppm) and $C_4$ (53 ppm) peaks for the -78° polymers, --which we
recall has an almost exclusively alternating structure--, we see
that they are clearly split, but by less than 1 ppm. We might at
first think this represents cis and trans structures. However,
experience with diene polymer spectra shows that when methylene
carbons are involved in a cis structure they shield each other by
8 to 10 ppm. This is due to the operation of the γ steric effect,
particularly strong when the carbon bonds actually eclipse each
other rather than being merely gauche. In chloroprene units one
expects the $C_4$ carbon to shift little between a cis and trans
structure because it always sees a bulky substituent across the
double bond. The $C_1$ carbon, however, sees such a substituent only
in the cis structure, in which it should accordingly be upfield
from the trans carbon. We believe the small peaks at 32 and 57
ppm represent the cis structure, as reflected in the $C_1$ resonances.
It can be clearly seen that trans strongly predominates. This
assignment is also confirmed by the trans C-C double bond stretch-
ing frequency at 1660 cm$^{-1}$; cis would be expected at 1652 cm$^{-1}$.

In Fig. 8 are shown the olefin carbon spectra of four copolymers as a function of temperature, including that shown in Fig. 6.
As we have seen, eight groups of resonances are recognizable.

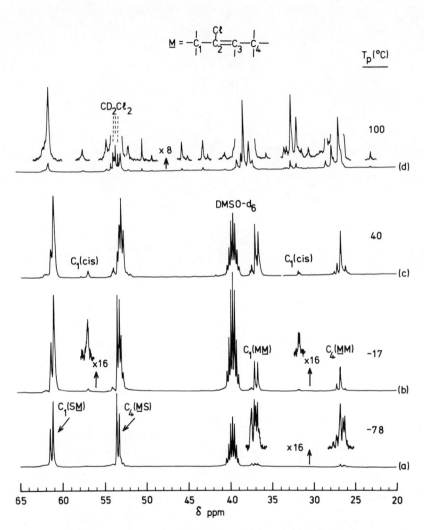

Figure 7.   A 90-MHz spectra of methylene carbons in four copolymers prepared at different temperatures. R Values are (top to bottom) 5.08, 1.64, 1.20, and 1.15. The top spectrum was run in $CD_2Cl_2$ and the others in $d_6$-DMSO.

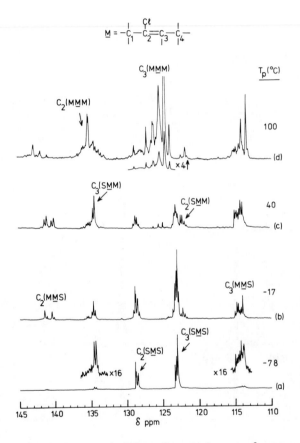

Figure 8.    A 90-MHz spectra of olefin carbons in four copolymers prepared at different temperatures.  R values are (top to bottom) 5.08, 1.64, 1.20, and 1.15. The top spectrum was run in $CD_2Cl_2$ and the others in $d_6$-DMSO.

This means we are dealing with four types of compositional triads (two types of carbons in each, $C_2$ and $C_3$): MMS, MMM, SMS, and SMM. The olefinic carbons look both ways along the chain and can tell what the neighboring monomer units are on both sides.

In the polymer prepared at -78°, the R value is nearly one and so almost the only sequence present is SMS. This enables this assignment to be made with confidence. The homopolymer type sequence, --MMM--, can be assigned by reference to the paper of Coleman, et al. (8) on polychloroprene, to which we have already made reference. The sequences MMS and SMM can be assigned as shown by appealing to chemical shift rules similar to those we have already discussed for polystyrene sulfone:

(a) A sulfone in the β position to a carbon exerts a shielding effect of -11 ppm, owing to the operation of the oxygen "γ-effect".

(b) A sulfone in the γ-position to a carbon exerts a deshielding effect of +8 ppm, arising from what may be termed an oxygen "δ-effect".

We then expect $C_2$ in MMS to be 11 ppm downfield from $C_2$ in SMS. Similarly $C_3$ in MMS should be 8 ppm upfield from $C_3$ in SMS. There are peaks in the expected positions. Assignments for SMM follow by default.

Sequence assignments in the methylene region (Fig. 7) are made considerably simpler by the fact that only compositional dyads rather than triads can be discriminated here. The $C_1$ carbons of the central chloroprene unit can distinguish units to the left but cannot distinguish those to the right, which are too far away. Similarly, $C_4$ carbons can distinguish units to the right but not to the left. The upfield pair of $C_1$ and $C_4$ resonances are in the positions corresponding to chloroprene homopolymers, whereas the downfield pair of resonances represent $C_1$ and $C_4$ carbons next to sulfone groups. Again, assignments are aided by recalling that alternating sequences predominate in the low temperature copolymers.

We note minor but complex resonances in both the olefinic and methylene spectra of the -78° temperature polymer. These arise in part from residual MM sequences but mainly from admixture of polymer formed at higher temperatures in the course of working up the product. We may note also in Fig. 7 that the proportion of cis 1,4 structures, as seen in the $C_1$ resonances, increases with polymerization temperature.

Let us finally consider the last form of isomerism with which we must deal. Geometrical isomerism and monomer sequence structure appear to be understood. However, there are clearly still splittings in both the olefinic and methylene spectra that have not been accounted for. This is evident even in the otherwise quite regular alternating structure of the -78° polymer, and must be due to head-to-tail head-to-head isomerism of the chloroprene units in the manner illustrated earlier. It is known that substantial proportions of head-to-head tail-to-tail units occur in

polychloroprene. The $C_1$ methylene carbons in SM dyads seem to
reflect only this form of isomerism, being unresponsive to others.
As we look at the behavior of this resonance as a function of
temperature, we note a surprising thing: as the polymerization
temperature increases it goes from a doublet to a singlet (Fig. 7).
Thus, in this respect, but in no other, the chain becomes more
regular with increasing temperature rather than less.  It has so
far been universally observed that polymer microstructure behaves
the opposite way, becoming more irregular with increasing temperature.  Why the difference in this case?  We believe that the
possibility of depropagation reactions explains this behavior.  At
low temperatures, forward propagation is favored; both head-to-
tail and head-to-head propagation are nearly equally favored,
head-to-tail being slightly preferred.  This is not really the
thermodynamically preferred ratio of structures, which is evidently almost exclusively head-to-tail, but nothing can be done to
correct this as the structure is, so to speak, trapped.  As the
temperature is increased, however, the depropagation reaction
comes more and more into play.  A monomer which comes in the
"wrong" way, i.e. head-to-head, gets a further chance to go out
and come back in again.

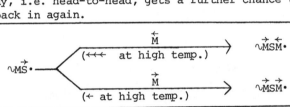

The depropagation reaction at the top is favored over that at
the bottom.  Note that this explanation applies only to MSM sequences.  Such a corrective device is not open to MM sequences,
in which we may expect the fraction of head-to-head to increase
with temperature in the "normal" manner.

I have discussed two cases of chain microstructure determination in $SO_2$ copolymers.  First, the styrene-$SO_2$ system, which
exhibits the general kinetic and compositional behavior of such
systems, particularly as a function of polymerization temperature.
Second, and considerably more complicated, is the chloroprene-$SO_2$
system.  This one represents the limit of what can be handled and
more or less completely solved at the present time.  To do so
requires about all our resources: [13]C at superconducting frequencies and a variety of strategies for carbon type assignment,
compositional assignment, and determination of the mode of addition of the chloroprene units.

Polysulfone Chain Dynamics.  Carbon-13 nmr has been used recently to resolve a puzzling problem in the dynamics of polysulfone chains (9,10,11). Some years ago, Bates, Ivin and Williams (12)
reported that measurements of the dielectric dispersion in solutions of alternating 1:1 copolymers of sulfur dioxide with hexene-1
and 2-methylpentene-1 show no loss in the high frequency region, i.e.

beyond 1 MHz. A high frequency loss region is to be expected for
flexible polymers having electric dipole components perpendicular
to the main chain direction. It was further found that the frequency of maximum loss was inversely proportional to the degree of
polymerization raised to the power 1.5-2.0, being for example
about 24 kHz in benzene at 25° for polyhexene-1 sulfone having a
number average molecular weight of 210,000. From these observations it was concluded that the presence of sulfone groups so
stiffens the chain that overall tumbling is the only motion effective in relaxing the molecule in an oscillating electric field.
This conclusion is quite surprising as molecular models and comparison to other polymers of comparable glass temperature (polystyrene and polymethyl methacrylate) do not suggest such a very
high degree of steric hindrance. Carbon-13 relaxation measurements
enable one to obtain at least an approximate measure of the chain
flexibility in such polymers. If we assume that a single correlation time $\tau_C$ describes the motion of the backbone chain and that
this motion is isotropic, then the following equations apply ([13]):

$$\frac{1}{NT_1} = \frac{1}{10} \frac{\gamma_H^2 \gamma_C^2 \hbar^2}{r_{C-H}^2} \cdot \chi(\tau_c)$$

For the nuclear Overhauser enhancement:

$$\eta = \frac{\gamma_H}{\gamma_C} \frac{6\tau_c \big/ \left[1+(\omega_H+\omega_C)^2\tau_c^2\right] - \tau_c \big/ \left[1+(\omega_H-\omega_C)^2\tau_c^2\right]}{\chi(\tau_c)}$$

where

$$\chi(\tau_c) = \frac{\tau_c}{1+(\omega_H-\omega_C)^2\tau_c^2} + \frac{3\tau_c}{1+\omega_C^2\tau_c^2} + \frac{6\tau_c}{1+(\omega_H+\omega_C)^2\tau_c^2}$$

The symbols have their usual meanings ([13]). From measured values
of $NT_1$ and $\eta$ on a poly(butene-1 sulfone) of degree of polymerization 700 the values of $\tau_c$ (in nanosec.) shown in Table III are
obtained. The discrepancy between the values of $\tau_c$ from $NT_1$ and
from $\eta$, particularly marked for the side-chain motions, indicates
the inadequacy of the single-$\tau_c$ model. Nevertheless it is evident that the backbone motions are relatively rapid. (Comparison
to polybutene-1 ([14]) shows that $SO_2$ groups retard the motion of
the copolymer chains by a factor of about 50.) The question now
becomes: why are these rapid motions NMR-active but dielectrically
inactive? One possible type of motion which would account for
this is shown in Fig. 9. Five backbone bonds and six main-chain
atoms are involved, i.e. the sequence C-S-C-C-S-C, with concerted
segmental transitions about two C-S bond, allowing interconversion
of the ttt, $g^+tg^-$, and $g^-tg^+$ conformers. The backbone C-C bond remains trans; the sulfone dipoles remain antiparallel and cancel
each other.

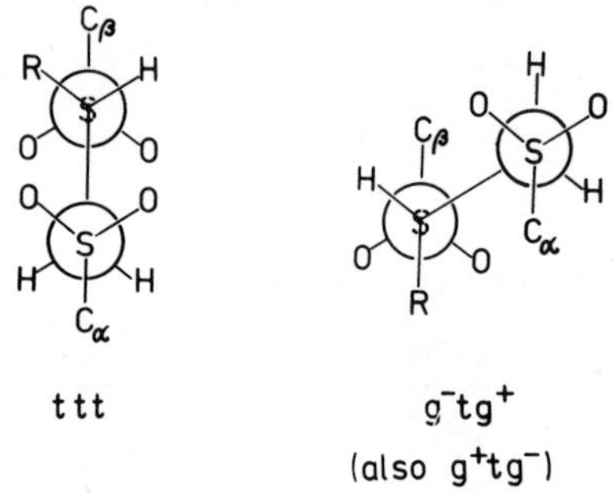

*Figure 9. Proposed allowed equilibrium conformational states for poly (α-olefin sulfones) in solution. Note that the sulfone dipoles cancel and that during the transitions* ttt ⇄ g⁻tg ⇄ g⁺tg⁻ *there is no net reorientation of these dipoles (dielectrically inactive motions), but there is a reorientation of backbone C–H vectors (C-13 NMR active motions).*

Table 3

Correlation Times for Backbone and Side-Chain Motions
in Poly(but-1-ene sulfone) of $P_n$ = 700 as a 25% w/v
Solution in Chloroform-d, Deduced from the Simple
Isotropic Single-$\tau_c$ Motional Model

| Temp, | Backbone from | | CH$_2$ from | | CH$_3$ from | |
|---|---|---|---|---|---|---|
| | | | Ethyl Branch | | | |
| K | NT$_1$ | $\eta$ | NT$_1$ | $\eta$ | NT$_1$ | $\eta$ |
| 328 | 22 | 2.6 | 0.29 | 1.9 | 0.027 | 1.3 |
| 313 | 23 | 3.1 | 0.33 | 2.2 | 0.030 | 1.3 |
| 298 | 27 | 3.4 | 0.38 | 2.7 | 0.039 | 1.3 |

Somewhat surprisingly, a 2:1 styrene:SO$_2$ copolymer is slightly more flexible than polystyrene itself (9). The styrene-
styrene (SMMS) units evidently constitute flexible joints which
permit an overall freedom of motion despite local constraints at
SMS units. The greater length of the C-S bond compared to that of
the C-C bond (0.180 vs. 0.154 nm) may also contribute.

LITERATURE CITED

1.  Solonina, W., Zh. russk. fiz.-khim. Obshch., (1898), 30, 826.
2.  Solonina, W., Chem. Zent., (1899), 1, 249.
3.  Dainton, F. S. and Ivin, K. J., Nature, (1948), 162, 705.
4.  Barb, W. G., Proc. Roy. Soc., London, Ser. A, (1952), 212, 177.
5.  Matsuda, M., Iino, M., and Tokura, N., Makromol. Chem., (1962),
    52, 98.
6.  Tokura, N. and Matsuda, M., Kogyo Kagaku Zasshi, (1961), 64,
    50.
7.  Cais, R. E., O'Donnell, J. H., and Bovey, F. A., Macromole-
    cules, (1977), 10, 254.
8.  Coleman, M. M., Tabb, D. L. and Brame, E. G., Rubber Chem.
    Tech., (1977), 50, 49.
9.  Cais, R. E. and Bovey, F. A., Macromolecules, (1977), 10, 757.
10. Stockmayer, W. H., Jones, A. A., and Treadwell, T. L.,
    Macromolecules, (1977), 10, 762.
11. Fawcett, A. H., Heatley, F., Ivin, K. J., Stewart, C. D. and
    Watt, P., Macromolecules, (1977), 10, 765.
12. Bates, T. W., Ivin, K. J. and Williams, G., Trans. Faraday
    Soc., (1967), 63, 1964.
13. See, for example, Shaefer, J., "Structural Studies of Macro-
    molecules by Spectroscopic Methods", K. J. Ivin, Ed., Wiley-
    Interscience, New York, NY, 1976, pp. 20.-226.
14. Schilling, F. C., Cais, R. E., and Bovey, F. A., Macromole-
    cules, (1978), 11, 325.

## Discussion

D.J. Worsfold, NRC, Ont.  Problems often arise in carbon-13 NMR work with the integration of the peaks.  In the analysis of copolymer systems it would be useful to have very good peak integration.  Can you in fact succeed in this?

F.A. Bovey.  That is a question I wish Dr. Cais were here to answer.  The assumption generally in such problems is that the Overhauser enhancement is at least the same for all peaks. I am not certain this is a justifiable assumption.  In fact, if you study Table II carefully you will see Overhauser enhancements which are in fact not necessarily the same.  That this can lead to problems is quite clear.  If work is done at a sufficiently high temperature I think the assumption of equal Overhauser enhancements is probably reasonably valid.  In this case the peaks can be intercompared.  We usually do this by old fashioned planimeter methods rather than by using the integrator on the instrument.  It seems to be more reliable.  This matter of integration is a definite problem which is becoming increasingly recognized.

A.A. Jones, Clark Univ., Mass.  I would tend to agree with the dynamic picture you presented.  The idea that the high frequency motion in the sulfone hexene type polymers doesn't cause a net relaxation of dipoles but moves the magnetic dipole-dipole interaction around to cause nuclear relaxation is, I think, the crux of the matter.

G. Babbitt, Allied Chemical Corp., N.J.  You showed quite large shielding and deshielding effects of $SO_2$ units in the styrene-$SO_2$ copolymer chains.  They are very dramatic.  Have you made any efforts to explain this?

F.A. Bovey.  No, I think the first thing we would like to understand is the carbon carbon gamma or gauche effect from which we might proceed to understand hetero atom gauche effects. I do not think anyone understands either one of them right now -- at least I don't.  Most of the theoretical efforts which I have seen are not very convincing.  The original explanation of Cheney and Grant (B.V. Cheney and D.M. Grant, J. Am. Chem. Soc., 89, 5319 (1967) required the carbons to have protons on them in order for this effect to be evident.  This is certainly not true.  There are gamma effects for all kinds of atoms other than carbon which do not have attached protons.  These systems exhibit strong effects.  Even with the protons attached I never quite understood what their explanation was.  In fact it cannot be limited to that explanation because it can be seen in many situations.  I think every atom can have a gamma shielding effect.  Even fluorines have a gamma effect, as we

found in interpreting the C-13 spectrum of polyvinylidene
fluoride.   It will take a very potent theoretical person to
work out the explanation of these effects and this has not
happened yet to my knowledge.   Perhaps someone else would care
to comment on this question.

C.J. Carman, B.F. Goodrich, Ohio.   I want to substantiate
a point with regard to the hetero atom effect.   We have tried
to extract the gamma effect of chlorine atoms in PVC and
chlorinated polymers and we have found that it's not quite as
large as with oxygen and not necessarily predictable as to its
direction.   I think that sometime in the future the hetero atom
effects on gamma may be extractable and of interest.   They are
not very predictable as far as transfering from one system to
another.   They can be very large as with oxygen.

F.A. Bovey.   Alan Tonelli in our laboratory has been working
on the prediction of carbon chemical shifts in polymer chains and
he finds what you found for the polyvinyl chloride.   I think you
first pointed out that the gamma effect can be used to explain
such chemical shifts.   Polyvinyl chloride can be explained if
enough variable parameters are used.   PVC is not the simplest
case.   Polypropylene on the other hand can be explained with a
single constant but PVC apparently cannot.   Dr. Tonelli hasn't
published any of this yet;he is still working it out.

T.K. Wu, E.I. duPont de Nemours, Delaware.   Were your
spectra obtained at a field higher than 22.6 MHz?

F.A. Bovey.   I'm glad you asked that question.   I should
have pointed out very clearly that all of the spectra I showed
for the chloroprene system were done at 90 MHz while those of
the styrene system were obtained at 25 MHz.

T.K. Wu, E.I. duPont de Nemours, Delaware.   A couple of
years ago Dr. Obiano, Dr. Haine and I studied a simpler alternating copolymer, that is, theylene CO, which we called a semi-
alternating sopolymer because the ethylene content was variable.
From the C-13 and proton NMR spectra we felt we could treat this
copolymer as a hypothetical copolymer of ethylene and a ficti-
tious comonomer ethylene–CO.   Using that picture we found the
comonomer distribution was very close to a Bernoullian distribution.

RECEIVED March 13, 1979.

# Polysaccharide Branching and Carbon-13 NMR

F. R. SEYMOUR

Baylor College of Medicine, Texas Medical Center, Houston, TX 77030

A fortunate relationship exists between current C-13 n.m.r. spectroscopy and polysaccharide structure. The structures of many polysaccharides are complex, providing a variety of resonances resulting from different structural features. At 25 MHz, a typical currently available field-strength, many resonances are well resolved and this encourages attempts to assign these resonances to specific carbon atom positions. Unfortunately, C-13 n.m.r. is not a fundamental structural technique -- that is, at our present state of knowledge it is not possible to directly interpret a saccharide C-13 n.m.r. spectrum without recourse to comparison spectra of known compounds. However, the above disadvantage is redressed in that C-13 n.m.r. spectra allow an intimate view of molecular structure, especially with regard to the nature of the anomeric linkages for specific types of residues. This very precise anomeric linkage data complements our basic structural analysis technique, permethylation g.l.c.-m.s., a technique that provides no information about the anomeric configuration of the various linkages. The general structural analytical approach which has been developed for polysaccharide structural analysis has previously been discussed in this Symposium Series (1). In general, data has been taken from fragmentation g.l.c.--m.s., permethylation fragmentation g.l.c.--m.s. (2-5), high-pressure liquid-chromatography of acetolysis products (3), C-13 n.m.r. (6,7), and P-31 n.m.r. (8), with these data then being inter-related. The following discussion will primarily deal with C-13 n.m.r., but the additional data is important. For example, permethylation fragmentation g.l.c.--m.s. analysis, which has been performed for all polysaccharides discussed, provides definitive evidence for the type and degree of saccharide branching. All of the following C-13 n.m.r. spectra-to-structure correlations are in agreement with this additional data.

As the basic spectral analysis approach consists of comparing the C-13 n.m.r. spectra of a large number of polysaccharides of known, or partially known, structure, the availability of a

0-8412-0505-1/79/47-103-027$06.25/0

comprehensive polymer collection is important. Furthermore, in comparison to other structural techniques (e.g. g.l.c.--m.s. analysis) Fourier transform C-13 n.m.r. is relatively insensitive, and requires large amounts of material ($\sim$200 mg) for complete analysis at high signal-to-noise ratio in reasonable spectrometer acquisition times. When considering the total spectrometer time required for obtaining a large reference collection of spectra, the available sample size becomes critical. Extracellular microbial polysaccharides provide an excellent source of reference materials, available in relatively large amounts, and are known to contain a variety of structural features. Many polysaccharides consist of hexose residues linked in the pyranoside ring form. In addition, many members of this polymer class are exclusively composed of D-glucopyranosyl residues (the D-glucans). Dextrans are an extensively studied sub-group of D-glucans, and are polymers which contain a large percentage of (1→6)-linked α-D-glucopyranosyl residues. Our original C-13 n.m.r. spectral studies were carried out on dextrans for four reasons: a) the compounds were available in suitable amounts, b) an abundant literature on the characterization of these compounds existed (9, 10), c) we had just completed accurate g.l.c.--m.s. structural analysis for many of these compounds (which indicated the different polymers had varied and well-defined structures), and d) it was hoped that the relatively small structural differences in various dextran series would allow a precise correlation between changes in structure to changes in spectra. Once the general spectral relationships were established for the various dextrans, it would then be possible to apply these relationships to polysaccharides of more diverse character.

It should be emphasized that although many structural features of the various polysaccharides are known, in general these polysaccharides are by no means completely characterized. However, with extensive C-13 n.m.r. chemical shift and intensity data, it is possible to employ these reference spectra to further characterize the compounds under consideration.

An important contribution to the analysis of 25 MHz spectra was the realization that elevated sample temperatures greatly aid spectral acquisition and interpretation (6). On raising the sample temperature the C-13 resonances narrow, providing both enhanced resolution and decreasing (by about 1/4) the number of acquisitions necessary to obtain a comparable signal-to-noise ratio for a corresponding ambient spectrum. It is for this reason that we normally record and reference our polysaccharide spectra at the highest temperature (90°) readily compatible with an aqueous solution. However, such high-temperature spectra must be interpreted with care due to the chemical shift dependence on temperature ($\Delta\delta/\Delta T$) for each resonance involved. In general, all 90° resonances are displaced down-field (relative to ambient, 34°, resonances) by $\sim$1 p.p.m. -- when referenced against the deuterium lock. The $\Delta\delta/\Delta T$ for different resonances vary as much as two

fold. Within spectrometer error $\Delta\delta/\Delta T$ is constant for dextrans
through out the 30--90° range. Though the individual resonances
narrow at elevated temperatures, the general "profile" of the
spectrum is constant for most polysaccharides that have been
studied (8, 12) -- the few exceptions will be discussed below.

Our original approach to polysaccharide C-13 n.m.r. spectral
analysis consisted of making a minimum number of hypotheses about
expected structure-to-spectra relationships (8). By then
comparing spectra to known structure for a series of D-glucans, we
attempted to establish the validity of these hypotheses and to
establish how diverse a structural difference could be accommodated
The hypotheses were as follows. Firstly, that each polymer could
be considered as an assembly of independent saccharide monomers.
Secondly, that these hypothetical saccharide monomers would be O-
alkylated (O-methylated) in the same positions as the actual
saccharide linked residues (it had previously been established
that O-methylation of any α-D-glucopyranosyl carbon atom position
resulted in a down-field displacement of ∿10 p.p.m. for the
associated resonance). Thirdly, that each differently substituted
residue would have a completely different set of chemical shift
values for each carbon atom position (different from the unsubstituted saccharide) but that only the carbon atom positions
involved in inter-saccharide linkages would have $\Delta\delta$ greater that
1 p.p.m. And, fourthly, that the hypothetical O-alkylated
residues would contribute resonances to the total spectrum proportional to their mole ratio in the polymers.

At this point it is convenient to reiterate the major
features and relationships normally encountered for polysaccharide
structures. Each saccharide residue is normally linked through
the reducing (hemi-acetyl) functional group (the anomeric position)
to the hydroxy position of another residue. Each residue contains
a number of hydroxy positions, but only one anomeric position.
When a residue is linked only through the anomeric position, then
that residue is a non-reducing terminal residue. When a residue
is linked through both a hydroxy position  and an anomeric
position, then a linear, chain extending residue results. In like
manner, should a residue be linked through both the anomeric
position and two hydroxy positions, then a branch-point residue
results. In general, any polysaccharide contains a single
reducing end group and b+1 terminal residues, when b represents
the number of branch-point residues. Therefore, large polysaccharides effectively contain no reducing end groups, and an
equal number of branch-point residues to terminal non-reducing
residues.

Dextrans

As previously stated, our C-13 n.m.r. spectral analysis was
first applied to dextrans. In the spectral analysis hypotheses
described above, it is an inherent assumption that each residue

will be considered independently, without regard to the nature of
the adjacent linked residues.  This procedure was employed for
simplicity of spectral analysis, and on the assumption that such
adjacent saccharide effects would be undetectably small.  However,
we knew that such adjacent saccharide effects would ultimately
need to be considered.  Our own P-31 n.m.r. spectra (8) had shown
such effects-at-a-distance, and a previously published pullulan
spectrum had shown resonance splitting which had convincingly
been explained as due to changes in (1→4)-linked α-D-glucopyranosyl residues which come before and after a (1→6)-linked
α-D-glucopyranosyl residue in the linear chain (12).  However, a
large amount of data accumulated for the dextrans over the past
three decades indicate that these compounds contain a backbone
composed exclusively of (1→6)-linked α-D-glucopyranosyl residues,
with branching occasionally occurring from this backbone chain
via the 2, 3, or 4-hydroxy positions (9).  Such branching effects
need not introduce as profound a conformational change into the
individual residues as does the introduction of a different
linkage type into the polysaccharide backbone.  In practical terms
the hypothesis of isolated residues has served very well for the
analysis of the dextran spectra.  Unsubstituted α-D-glucopyranoside (and apparently any hexapyranoside) exhibits six resonances
which can be observed in three regions:  the 90--95-p.p.m. region,
containing the anomeric (C-1) carbon resonance; the ∿70--75-p.p.m.
region, containing the C-2, C-3, C-4, and C-5 resonances; and the
60--70-p.p.m. region containing the C-6 resonance. On O-substitution the resonance of the anomeric carbon is displaced down-field
by ∿10 p.p.m. to the 97--102-p.p.m. region.  Similar O-substitution of carbon atoms with resonances normally in the 70--75-p.p.m.
region results in displacements into the 80--85-p.p.m. region.
The free C-6 resonance, at ∿68 p.p.m., on O-substitution is
displaced to ∿62 p.p.m., so that both free and linked carbon
resonances are in the 60--70-p.p.m. region.  Therefore, as the
percentage of free (reducing) D-glucose residues contained in the
dextrans is insignificant, only four spectral regions are of
interest:  the 60--70-p.p.m. region, containing both free and
linked C-6 resonances, the 70--75-p.p.m. region containing the
free C-2, C-3, and C-4 resonances and the pyranoside ring C-5
resonance; the 75--85-p.p.m. region containing linked C-2, C-3,
and C-4 carbon resonances, and the 95--103-p.p.m. region containing linked anomeric (C-1) resonances (see Figure 1, upper
spectrum).

   The two spectral regions most amenable to analysis are the
95--103-p.p.m. region (C-1 resonances) and the 75--85-p.p.m.
region (the linked C-2, C-3, and C-4 resonance position).  This
phenomenon could easily be expected, as these two regions contain
the resonances of the carbon atoms which are directly participating
in the inter-saccharide linkages.  The 75--85-p.p.m. region
presents a rather straight-forward situation, as each branch-type
results in a specific branching residue (e.g. 2-O-branching from

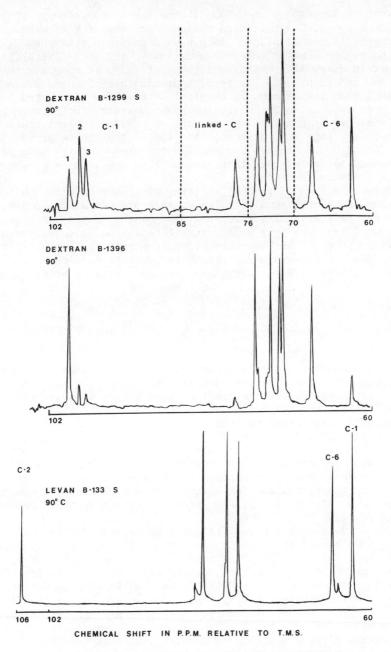

*Figure 1. The 90° C-13 NMR spectra for: (top) dextran B-1299 Fraction **S**; (middle) dextran B-1396; and (bottom) levan B-133 Fraction **S**.*

a backbone composed of 6-O-substituted α-D-glucopyranosyl residues
yields a 2,6-di-O-substituted residue) and the linked carbon
resonance is now displaced ∿10 p.p.m. down-field from the 70--75-
p.p.m. region.  At 90° α-(1→2)-, α-(1→3)-, and α-(1→4)-linkages
result in diagnostic resonances, respectively, at approximately
78, 80, and 82 p.p.m.  The term "approximately" is used advisedly
because some differences in chemical shifts have been noted for
the same type of inter-sugar linkages.  However, if the spectra
are analyzed by considering the type of residue present, rather
than the linkage type present, then the diagnostic resonances in
the 75--85-p.p.m. region are consistent.  Two examples will be
presented to demonstrate this residue type specificity effect for
the diagnostic resonances.

Several series of graded (in terms of degree of branching)
dextrans have been studied.  The example presented here deals
with dextrans branching through 2,6-di-O-substituted α-D-glucopyranosyl residues (12).  The average repeating unit for these
dextrans is shown in Structure 1, where G represents the D-glucopyranosyl residue and n represents the average number of
6-O-substituted α-D-glucopyranosyl residues between branch points.

$$-\{-\alpha-(1\to6)-G-[-\alpha-(1\to6)-G-]_n\}_x-$$
$$\phantom{xxxxxx}|$$
$$G-\alpha-(1\to2)$$

*Structure 1*

Assuming that all α-(1→6)-linkages are in the dextran backbone,
our permethylation g.l.c.-m.s. data (which indicate that only the
residues shown above are present in this series of dextrans) are
consistent with a polymeric structure that is comb-like with side
branches a single residue long.

Figure 2 shows a specific example of the effect on a C-13
n.m.r. spectrum when the sample temperature is changed.  With
increasing temperature the resonances are displaced down-field and
become more narrow.  The upper plot of Figure 2 shows the linear
$\Delta\delta/\Delta T$ for the resonances designated in the spectra.  The dextran
anomeric $\Delta T$ effects are even greater than the selected example
shown in Figure 2.

The dextran series B-1299 fraction S, B-1402, B-1424, B-1422,
and B-1396 have been shown to have structure 1 with n respectively
equal to 1, 2, 3, 6, and 8.  These dextran (and the following
mannan and levan) designation numbers refer to the Northern
Regional Research Center (NRRL) designation for the producing
microorganism strain.  A single sharp resonance in the 75--85-
p.p.m. region was observed for each of the dextrans (see Figure 1,
upper and middle spectra), respectively at 77.79, 77.80, 77.79,
77.74, and 77.76 p.p.m.  These chemical shift differences
essentially represent the limits of our spectrometer error.

Comparison of two series of dextran data shows the importance
of considering the spectra in terms of specific residue type.  One

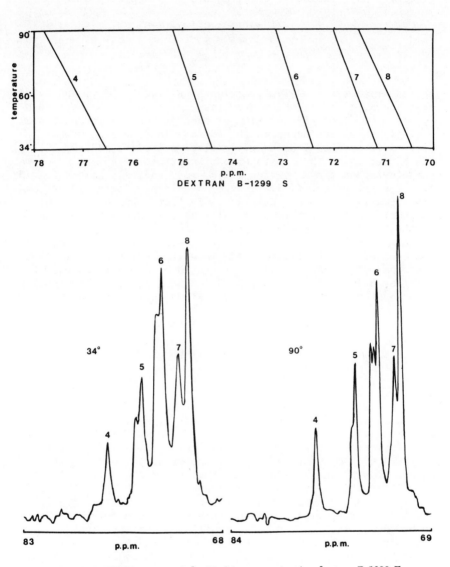

*Figure 2.  C-13 NMR spectra of the 69–84 ppm region for dextran B-1299 Fraction S (total 90° spectrum in Figure 1) at 34° and 90°. Top plot shows the Δ∂/ΔT for five resonances identified in the bottom spectra.*

dextran series is essentially the same as the above 2-O-sub-
stituted series, except the branching is through the 3,6-di-O-
glucopyranosyl residue (see Structure 2).  Representative of this

$$-\{-\alpha-(1\rightarrow6)-G-[-\alpha-(1\rightarrow6)-G-]_{\underline{n}}\}_{\underline{x}}-$$
$$G-\alpha-(1\rightarrow3)$$

*Structure 2*

series are dextrans B-742 fraction S, B-1142, B-1191, and B-1351
fraction S, dextrans which have respective values of $\underline{n}$ to be 0, 1,
2, and ⌄10.  The C-13 spectrum of each of these dextrans contains
a single sharp resonance in the 75--85-p.p.m. region with
diagnostic chemical shifts respectively of 82.81, 82.89, 82.84,
and 82.86; again the limits of our spectrometer accuracy.  However,
permethylation fragmentation g.l.c.-m.s. analysis has indicated
that several of the (1→3)-linkage containing "dextrans" differ
dramatically from other dextrans.  In fact, though these polysaccharides have been traditionally classed with the dextrans due
to their origin, they quite possibly fail to fulfill the
definition of "dextran" by not having backbones exclusively
composed of 6-O-substituted α-D-glucopyranosyl residues.  This
class of polysaccharides is represented by the S fractions of
dextrans B-1355, B-1492, and B-1501, and a general structure is
proposed for this class (structure 3) which is based on C-13 n.m.r.

$$-\{[-(1\rightarrow6)-G-(1\rightarrow3)-G-]_{\underline{n}}-(1\rightarrow6)-G-(1\rightarrow3)-G-\}_{\underline{x}}-$$
$$G-[-(1\rightarrow6)-G-]_{\underline{p}}-(1\rightarrow6)$$

*Structure 3 (all linkages α)*

and g.l.c.-m.s. data (13).
     The feature of immediate interest in the above general
structure is that relatively large (and for dextran B-1355
fraction S, very large) percentages of 3-O-substituted α-D-glucopyranosyl residues are present with relatively small proportions
of the 3,6-di-O-substituted α-D-glucopyranosyl residues.  Careful
examination of the 75--85-p.p.m. region for the above polymers
shows that the major resonance is now respectively at 83.31,
83.31, and 83.32 p.p.m.  These values are again indistinguishable,
and are clearly displaced from the diagnostic chemical shifts
observed for the dextran B-742 fraction S series, which
exclusively contain (1→3)-linkages associated with 3,6-di-O-
substituted α-D-glucopyranosyl residues.  In addition, dextran
B-1501 fraction S contains a relatively modest excess of 3-O-
substituted α-D-glucopyranosyl residues when compared with the
3,6-di-O-substituted α-D-glucopyranosyl residues, and careful
examination of the 75--85-p.p.m. region of this dextran displays
a major resonance at 83.32 p.p.m. (indicative of the former
residue and designated $\underline{d}$ in Figure 3, upper spectrum)

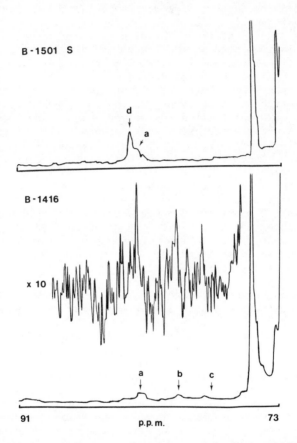

*Figure 3.  The 73–91 ppm 90° C-13 NMR region for the dextran B-1501 Fraction S and the dextran B-1416 spectra.  The letters identify: Resonance a (from the 3,6-di-0-substituted α-D-glucopyranosyl residue); Resonance b (from the 4,6-di-0-substituted α-D-glucopyranosyl residue); Position c (for the 2,6-di-0-α-D-glucopyranosyl residue); and Resonance d (from the 3-0-substiuted α-D-glucopyranosyl residue).  A tenfold scale expansion plot of dextran B-1416 is also shown.*

and a shoulder at 82.9 p.p.m. (indicative of the latter residue and
designated a in Figure 3, upper spectrum). In conjunction with
similar studies for (1→4)-linkage containing dextrans, it is
concluded that resonances have been observed which are specific
for the 3-O-substituted, 4-O-substituted, 6-O-substituted,
2,6-di-O-substituted, 3,6-di-O-substituted, and 4,6-di-O-
substituted α-D-glucopyranosyl residues, and these diagnostic
resonances are specific for each residue type.

The separation of these 75--85-p.p.m. diagnostic branching
resonances is indicated in the lower part of Figure 3. No
completely linear microbial dextran is known, but dextran B-1416
is a dextran example with low degree of branching. It is one of
the few dextrans which have been identified as containing a
mixture of branch-type residues. Both the 3,6-di-O-substituted
and the 4,6-di-O-substituted residue resonances are weakly present.
A third, and unexplained, very weak resonance (78.3 p.p.m.) is
near, but clearly displaced from, the position c (77.8 p.p.m.)
which is diagnostic for the 2,6-di-O-substituted α-D-glucopyranosyl residue.

In general, the spectra of any dextran is a composite
consisting of the six major resonances associated with linear
dextran (the 6-O-substituted α-D-glucopyranosyl residue) and
minor resonances associated with the branching and corresponding
terminal residues. Due to the corresponding terminal α-D-glucopyranosyl residue associated with each branch-point residue, it
could be expected that each branching type could contribute twelve
(2x6) minor resonances to the dextran spectra, and that the
"minor" resonances would steadily become more intense (compared to
the linear dextran resonances) as the polymer becomes more highly
branched. We have not observed all of these expected additional
resonances, and assume that in each case certain of these
expected resonances are not resolved from the closely packed
70--75-p.p.m. region.

The anomeric, 95--110-p.p.m. region presents a second
complementary spectral region for structural analysis. In contrast
to the 75--85-p.p.m. region, the anomeric region presents two new
resonances for each branching type -- one resonance from the
branch-point residue and a second from the corresponding terminal
residue. Based on a number of polymer analyses, the C-1
resonances are the most sensitive to structural change, and at
this point there is failure for the hypothesis that neighboring
residues have no effect on a given residue type's resonances
(this is the only failure of this hypothesis which has been noted
for the dextrans). The anomeric spectral region of the dextran
series branching through the 2,6-di-O-substituted α-D-glucopyranosyl residue (see Structure 1 and Figure 1) displays two
branching resonances (97.37 and 98.22 p.p.m.), in addition to the
linear dextran anomeric resonance. One new additional
resonance belongs to the C-1 of the branch-point residue and the
other additional resonance corresponds to the C-1 terminal residue.

If the hypothesis stating that the resonances of each residue
type were independent of adjacent residue structures, then the
C-1 resonance of the terminal residue would be easy to identify
by comparing the additional branching anomeric resonances of
2-branching, 3-branching, and 4-branching, thereby identifying
the consistent minor resonance. This is not the case, as a
comparison of the series of dextrans of different branching type
demonstrates. The spectra of each branching type (see Figure 4,
top spectra) display two additional anomeric branching resonances,
but for each type of branching both minor resonances differ from
the minor resonances of the other branching types of dextran. The
actual assignment of these anomeric resonances will be discussed
below. In addition, a small but distinct chemical shift is noted
for the anomeric resonance of the 3,6-di-O-substituted residue
(100.88 p.p.m.) compared to the 3-O-substituted (101.07 p.p.m.)
residue.

   Most importantly, in comparing the C-13 n.m.r. spectra of the
dextran series branching through the 2,6-di-O-substituted α-D-
glucopyranosyl residue to the permethylation g.l.c.-m.s. data,
it was concluded that the intensity of all resonances present is,
to a first approximation, proportional to the number of atoms of
that carbon position present; and also, in general the peak
heights of the minor resonances (compared to the major resonances
of linear dextran) are proportional to the degree of branching.
This effect was specifically confirmed by comparing the ratio of
peak heights of a branching anomeric resonance (99.2 p.p.m.) and
the linear dextran resonance (98.7 p.p.m.) to permethylation
data (12). However, this effect can be shown for any well
resolved minor resonance of any of the different branching type
dextran series. The establishment of such relationships is of
importance as C-13 resonance intensities do not necessarily bear
any direct relationship to the percentage of specific position of
carbon atoms present.

   The general anomeric resonance C-13 n.m.r. region of D-glucans
is of considerable interest as it has been concluded, on the basis
of monomers, that β-configuration anomeric carbons have chemical
shifts down-field from α-configuration anomeric carbons. All
evidence in the literature at present confirms this general
observation, but until a wide variety of monomers and polymer
residues have been examined, it will be difficult to establish
exactly what the range of the α- and β-anomeric resonance regions
are, and whether these two regions can overlap. Currently it is
suggested that these regions (corrected to 90°) would be about
95--103-p.p.m. for the α-configuration, and about 103--108-p.p.m.
for the β-configuration. Due to large positive specific rotations,
the dextrans were known to be predominantly, if not exclusively,
α-linked. In none of our dextran spectra have we observed any
anomeric resonance down-field from 101.6 p.p.m., nor up-field from
97.2 p.p.m. This suggests that the dextrans contain no
(observable) minor β-linked residues, and that the general spectral

region for α-D-glucopyranosyl residues lies in the 97--102-p.p.m. region.

Relatively little spectral analysis has been done on the 60--70-p.p.m. or the 70--75-p.p.m. regions of the dextran spectra. The 70--75-p.p.m. region is closely packed with a number of resonances, and therefore the identification of specifically displaced carbon position resonances is difficult at our current spectral resolution. The 60--70-p.p.m. region provides a good indication of the degree of branching, by comparing the free C-6 resonance intensity (∿63 p.p.m.) to the linked C-6 resonance intensity (∿68 p.p.m.). However, little additional information is forthcoming as the C-6 position of the α-D-glucopyranosyl residue is relatively insensitive to the type of branching. In fact, the general insensitivity at the C-6 position can be observed in the relatively small Δδ of substitution (∿5 p.p.m.).

This C-13 n.m.r. analytical technique has been referenced against 27 dextrans which had been studied by periodate-oxidation and permethylation fragmentation g.l.c.-m.s. techniques (14). These three independent structural methods give data for all dextrans which are in accord with each other. The 27 dextrans studied represent 25 structues which differ in type or degree of branching. Such extensive cross-referencing of data give us increased confidence with regard to spectra-to-structure relationships.

The nature and relationships of the resonances in dextran spectra have been dealt with at length due to the wide variety of branching types available, and to the relatively subtle changes in structure available as one progresses through a specific dextran branching type series. The C-13 n.m.r. spectra of the following polysaccharides can now be dealt with in a more succinct manner by describing how they differ from the above observations.

## Synthetically branched amylose

This series of compounds, produced by adding D-glucopyranosyl monomers to amylose [the linear polymer composed of (1→4)-linked α-D-glucopyranosyl residues], was of interest as a number of different highly-branched comb-like amylose products had been produced by employing different types of condensation reaction conditions. In general, such D-glucopyranosyl additions result in the formation of β-linkages, but certain reaction conditions (specifically the Helferich reaction conditions) were anticipated to yield the addition of α-linked D-glucopyranosyl terminal residues to the linear amylose chain. Various techniques, including permethylation g.l.c. analysis, were able to demonstrate the degree of monomer incorporation and the position of attachment (at the C-6 position of the amylose residues). However, establishment of the anomeric configuration proved much more difficult, as most methods of anomeric linkage determination depend on bulk properties of the polymer (e.g. specific rotation) and also on the

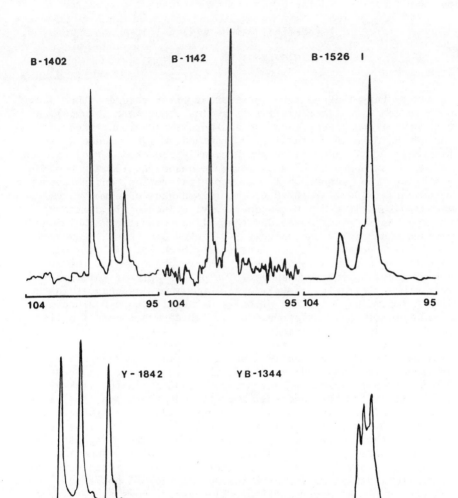

*Figure 4.   (Upper) Anomeric region of the C-13 NMR spectra at 90° for: dextran B-1402 (2,6-di-0-substituted branching); dextran B-1142 (3,6-di-0-substituted branching); dextran B-1142 (3,6-di-0-substituted branching); and dextran B-1526 Fraction I (4,6-di-0-substituted branching). For each dextran the most intense resonance (99.6 ppm) represents the 6-0-substituted α-D-glucopyranosyl residue. (Lower) Anomeric region of the 90° C-13 NMR spectra for mannan Y-1842 and the 76–90 ppm region for mannan YB-1344*

fact that the amylose backbone, now incorporated into the new polymer, contributed large amounts of α-linkages (see Structure 4).

$$-\{-\alpha-(1\to4)-G-[-\alpha-(1\to4)-G-]_{\underline{n}}\}_{\underline{x}}-$$
$$|$$
$$G-\beta-(1\to6)$$

*Structure 4*

To answer this linkage type question we first recorded the spectra of glycogen (a polymer similar to amylose, but more soluble), of both methyl α-D-glucopyranoside and methyl β-D-glucopyranoside, and of the unknown synthetically branched amylose products (see Figure 5). All examples of synthetically branched amylose proved to have a strong branching anomeric resonance at 104.4 p.p.m. (15) (well into the β-anomeric linkage spectral region) and no anomeric resonance at 100.1 or 101.6 p.p.m. (resonances observed for 4,6-di-O-substituted α-D-glucopyranosyl branching in dextrans). On this basis it was concluded that all condensation conditions yielded essentially exclusively β-D-linked branching. Furthermore, the total spectrum of the synthetically branched amyloses was essentially a composite of the amylose resonances plus the methyl β-D-glucopyranoside resonances. Two points were of interest in this study. Firstly, the glycogen diagnostic resonances in the 75--85-p.p.m. region (that of the 4-O-substituted α-D-gluco-pyranosyl residue) was at 79.4 p.p.m. which was somewhat different from the diagnostic branching resonance of 4-O-branched dextran at 80.4 p.p.m. (the 4,6-di-O-substituted α-D-glucopyranosyl residue). As was previously shown for α-(1→3)-linkages for dextran, this again indicates that the diagnostic resonances are dependent on the residue type present, and not on the linkage type. Secondly, resonances similar to both the above (1→4)-linked glycogen and dextran resonances are observed in the synthetically branched amylose. The presence of these diagnostic resonances in the synthetically branched amylose is not surprising as this compound contains both 4-O-substituted and 4,6-di-O-substituted α-D-glucopyranosyl residues. However, the latter residue is now linked to a β-D-glucopyranosyl residue at the C-6 position. The branched amylose spectrum (Figure 5) clearly shows these effects -- resonance a (80.3 p.p.m.) is essentially identical to the diagnostic resonance for dextrans branching through the 4,6-di-O-substituted α-D-glucopyranosyl residue; resonance b (79.1 p.p.m.) is identical to that for glycogen (and therefore linear amylose). and resonance c (77.9 p.p.m.) corresponds to the closely spaced doublet observed for methyl β-D-glucopyranoside. The corresponding methyl α-D-glucopyranoside has no resonances in the 70--75-p.p.m. region.

## Extracellular mannans

In contrast to the above data, which for dextrans and amyloses

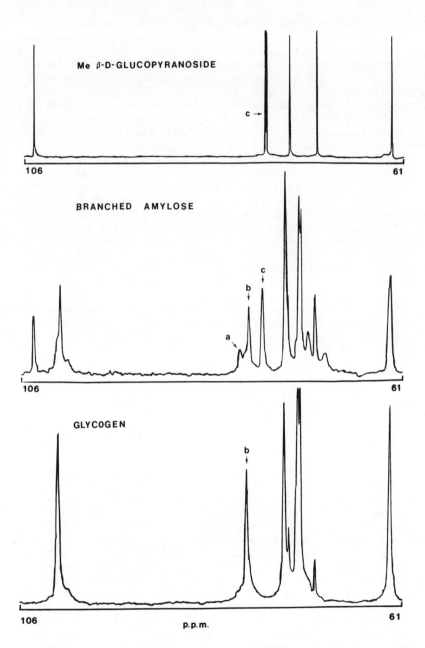

*Figure 5.   The 90° C-13 NMR spectra of methyl α-D-glucopyranoside (the aglycon resonance at 55 ppm is not shown), a branched amylose (sample OAG-II-3), and rabbit liver glycogen*

show a relatively simple structure-to-spectra relationship, the
mannan spectra present greater difficulties in interpretation.
This is quite possibly due to two different reasons.  On the basis
of permethylation g.l.c.-m.s. and h.p.l.c. acetolysis data the
various mannan structures appear to display greater structural
diversity than the dextrans (3).  In addition, fewer minor
structural variants (e.g. differences in degree of branching) are
available for these mannans.  In general, the mannans contain
greater percentages of non-(1→6)-linkages than do the dextrans.
Furthermore, permethylation g.l.c.-m.s. indicates that mannans, in
contrast to the dextrans, often have these non-(1→6)-linkages
associated with linear chain extending residues, rather than as
branch-points associated with branched residues also containing
(1→6)-linkages.  Acetolysis data also indicates that many of these
non-(1→6)-linked residues are sequentially linked.  In accord with
the dextran structures, we have previously proposed mannan
structures which contain all (1→6)-linkages in the backbone --
though this may well not be the case for all or any of these
polysaccharides.  Two typical mannan spectra will now be examined
to illustrate the differences between mannans and dextrans.

Mannan YB-1344 -- By combining the permethylation g.l.c.-m.s.
analysis data with the postulate that all (1→6)-linkages are in a
linear backbone, the following average repeating unit can be
constructed (see Structure 5), where M represents the D-manno-

$$-[-(1\to6)-M-]_x-$$
$$|$$
$$M-(1\to3)-M-(1\to3)-M-(1\to3)$$

*Structure 5 (all α-linkages)*

pyranosyl residue.  Such a structure differs radically from the
essentially linear dextrans' structure which contain single
residue branches.  However, comparison of the mannan 75--85-p.p.m.
diagnostic region to that of a typical (1→3)-linked dextran shows
a profound difference.  Where the dextrans display a single sharp
resonance, mannan YB-1344 displays a series of overlapping
resonances (see Figure 4).  With the exception of the anomeric
resonances, the total YB-1344 spectrum displays many more
resonances than a comparable (similar degree of branching) dextran
spectrum.  For some reason the individual 3-O-linked α-D-mannopyranosyl residues are in a greater variety of environments than
the dextran residues, and the logical inference from these data is
that such environmental differences result from side-chains of
greater length than a single saccharide residue.

Mannan Y-1842 -- Compared to previous dextran spectra-to-
structure relationships, this polymer displays an unexpected
spectrum.  Careful permethylation g.l.c.-m.s. data indicate a
2:1 ratio of 2-O-substituted D-mannopyranosyl residues to 3-O-
substituted D-mannopyranosyl residues, with very few branching
and terminal residues.  Such an assembly of residues yield the

following average repeating unit (Structure 6). The (1→2)- and
(1→3)-linkages have been shown in an ordered manner, but per-

$$-[-(1→2)-\alpha-M-(1→3)-\alpha-M-(1→2)-\alpha-M-]_x-$$
*Structure 6*

methylation g.l.c.-m.s. data actually yield no information on the
ordering of such residues. The permethylation g.l.c.-m.s. data
also provide no information about the anomeric linkages of the
residues. The spectra-to-structure relationship problem for this
polymer is obvious, and is described as follows. When the
hypotheses which were successfully applied to dextrans are applied
in this case, the residues present, known from permethylation
analysis, simply cannot give the observed spectrum. Two 2-O-
substituted α-D-mannopyranosyl residues and one 3-O-substituted
α-D-mannopyranosyl residue are expected to yield only two anomeric
resonances, with the intensity of the former residue's resonances
twice the intensity of the latter residue's resonances. Such an
effect is not observed for the anomeric region of the mannan
Y-1842 spectrum -- in fact three well spaced resonances of
approximately equal intensity are observed (see Figure 4). Either
the limiting situation for our working hypothesis has been
reached, or mannan Y-1842 does not have the general structure
we have assigned it. However, the combined data from permethylation fragmentation g.l.c.-m.s., specific rotations, and
saccharide surveys by g.l.c.-m.s. and paper chromatography all
support the presence of the specific type and ratio of saccharide
residues indicated in Structure 6.

If the above mannan structures are properly understood, then
the spectra-to-structure relationships must be reexamined. The
permethylation g.l.c.-m.s. data show that the structure of mannan
Y-1842 profoundly differs from the above dextran structures as
this mannan has non-(1→6)-linkages incorporated into the backbone.
Previous examination of pullulan, a D-glucan containing a 2:1
ratio of (1→4)- and (1→6)-linkages in the backbone, showed a
resonance splitting of both the C-1 and C-4 resonances originating
from the 4-O-substituted α-D-glucopyranosyl residues (but the C-1
resonance was split by only ∿0.5 p.p.m.). This spectrum was
convincingly interpreted by postulating pullulan to be a highly
ordered polysaccharide of the following structure, and assuming

$$-[-(1→4)-\alpha-G-(1→6)-\alpha-G-(1→4)-\alpha-G-]_x-$$
*Structure 7*

that those 4-O-substituted α-D-glucopyranosyl residues are in
different chemical environments and will have different chemical
shifts for various carbon atom positions.

Structures 6 and 7 are obviously very similar, both containing a 2:1 ratio of differently linked residues in a uniform
pattern. This then may provide the explanation for the three

equal intensity anomeric resonances of mannan Y-1842 via the
"before" and "after" 2-0-substituted α-D-glucopyranosyl residues.
It is inherent in this assumption that mannan Y-1842 is a highly
ordered linear polymer.  Such a postulated structure needs to
explain why the similar residue anomeric resonances of mannan
Y-1842 are so much more separated than those of pullulan, and
there are two possible explanations.  Mannan Y-1842 contains
linkages through the C-2 and C-3 positions, which are much closer
to the anomeric carbon than the C-4 and C-6 linkages of pullulan,
therefore, the effect observed for pullulan could be magnified
for mannan Y-1842.  In addition, the difference between α-D-glucopyranosyl and α-D-mannopyranosyl residues lies in the configuration of the C-2 position, with the α-D-mannopyranosyl containing
the C-2 linkage in a position axial to the ring and closer to
the C-1 position.  For the above reasons, the C-1 resonance of the
2-0-substituted α-D-mannopyranosyl position could be quite
dependent on the geometry of the adjacent linked saccharide.
     The general thrust of the above arguments can be summed as
follows.  The chemical shifts of a given residue type are
minimally affected by the nature of adjacent linked residues as
long as those adjacent residues are not part of the polymer
backbone or side-chain.  However, once such different linkages
are incorporated into a backbone chain, or short side-chain, then
the relative positions of the residues can result in a mutual
interaction displacing the chemical shifts associated with
specific carbon positions.  If these arguments are correct, they
then explain why mannan spectra are more complex (and making the
corresponding assignments is more complicated) than for the
dextrans and amylose derivatives.

## Levans and β-D-fructofuranosyl compounds

     Though the above polymers have shown a wide variety of
resonances and spectra-to-structure relationships, they have all
had in common structures composed exclusively of hexapyranosyl
residues.  The levans, composed of β-D-fructofuranosyl residues,
are of interest in that these polymers can be produced by the
same microorganisms which produce dextrans, and also of interest
for examining the structure-to-spectra relationships for this new
ring system.  The C-13 n.m.r. spectra of the α-D-glucopyranosyl
and β-D-fructofuranosyl residues have an unusual relationship, for
when the two spectra are compared there is essentially no overlap
of resonances (see Figure 1).  The resonances of dextran and
levan are remarkably evenly spaced across the 60--110-p.p.m.
region of the C-13 n.m.r. spectrum.  The lack of development of
g.l.c.-m.s. ketose techniques, comparable to those for aldose
residues, make the general approach to establishing the branch-
type data less feasible than for the dextrans and mannans.
Furthermore, as will be seen below, the currently isolated
β-D-fructofuranosyl polymers display much less variety of

structural features than do the D-glucans and D-mannans.  However, the above resonance spacing of the β-D-fructofuranosyl residues allows an alternative approach to establishment of spectra-to-structure relationships (16).

The fortunate existence of a group of oligomers, which contain the β-D-fructofuranosyl residue as a central moiety being O-substituted at various positions by the α-D-glucopyranoside group, allows the study of the displacement of specific resonances due to this substitution.  For example, 2-O-substitution of the β-D-fructofuranoside residue gives sucrose; 2,3-di-O-substitution gives melezitose.  Employing Δδ of substitution data, plus chemical shift data from the 1,2-di-O-substituted β-D-fructofuranosyl residues of inulin, allow resonance assignment to the specific carbon positions of the β-D-fructofuranosyl residue.  In general, it is concluded that O-substitution at the C-1 and C-6 positions of the β-D-fructofuranosyl residue results in Δδ displacements of only ∿3 p.p.m., which are much smaller than the average Δδ of O-substitution for the α-D-glucopyranosyl residue.  However, the C-1 and C-6 positions are primary alcohol positions out of the ring system, and a smaller than average Δδ of O-substitution was also observed for the α-D-glucopyranosyl C-6 resonances (which also is a primary alcohol position out of the ring).  In addition, the resonance of the C-2 (anomeric) position of the β-D-fructofuranosyl residue gives evidence of being much less sensitive to structural change than the corresponding C-1 (anomeric) position of the α-D-glucopyranosyl residue.

The immediate application of this levan C-13 n.m.r. approach has been to analyze the levan fractions resulting from acid hydrolysis of native levan, and to survey a group of levans from the NRRL collection which were produced by diverse bacterial strains.  We have concluded that acid hydrolysis has little effect on modifying the average degree of branching for resulting levan fragments.  The group of NRRL levans exhibited a very close similarity of structure (especially when compared to the wide diversity of structure encountered in the corresponding NRRL dextran collection), but do show certain distinct differences in minor resonances which suggest that these levans differ in type of branch-point residues.  The distinct differences in resonance position between dextran and levan C-13 n.m.r. spectra indicate that C-13 n.m.r. could be a very usable technique for quantitation of dextran-levan mixtures, and indeed it would appear that a large amount of structural evidence could be obtained for one class of these polymers in the presence of the other.

Resonance relaxation measurements

The above data have indicated that C-13 n.m.r. spectra can distinguish between subtle structural effects in polysaccharides, and that in many cases resonances can be assigned to specific

carbon atom positions. The information arising from the
assignment of resonance to carbon atom position can be very
important in establishing the effect of structural changes on
resonance displacements. If such effects were clearly understood, then a C-13 n.m.r. spectrum could be accurately predicted
for a given polymer, or conversely, an intimate knowledge of a
specific polymer could be established solely on the basis of a
C-13 n.m.r. spectrum. Furthermore, neither our g.l.c.-m.s. data,
nor a simple C-13 n.m.r. spectrum, yield any evidence about the
relative position of individual residues within a polymer.
Fortunately, relaxation data provide certain evidence for both of
these areas of uncertainty.

Analysis of the anomeric resonances produced due to branching
can be very frustrating when based on the simple C-13 n.m.r.
spectra. For the dextrans these anomeric branching resonances are
well resolved one from another, and from the linear polymer
resonance, but the introduction of a branch-point residue into
a linear polysaccharide results in the creation of an additional
terminal residue with its associated anomeric resonance. Therefore, each type of branching introduces two sets of new
resonances into the spectrum -- one set from the branch-point
residue, and one set from the terminal residue. This duplicate
set of residue resonances causes confusion in resonance assignment as there is no way to separate and assign these sets of new
resonances. As was previously noted, not all carbon atom
positions can be expected to yield resonances of equal intensity
in a given spectrum. The actual resonance intensity, and half-
height width, is dependent on relaxation times ($T_1$), and on
nuclear Overhauser effects (n.O.e.). The $T_1$ values for a specific
resonance is essentially a measure of the relative degree of
mobility within the polymer at the specific carbon atom position
related to that resonance. The n.O.e. for a given resonance
is a measure of carbon-proton dipole reorientation in competition with solvent interaction with a carbon atom associated
with that resonance. It is possible to also calculate $T_{1DD}$, a
function of $T_1$ and n.O.e., but for the polymers discussed here
the relative sets of values for $T_1$ and $T_{1DD}$ measurements parallel
each other. The application of these concepts will be
demonstrated by the following examples.

Dextran B-1299 fraction S yields a simple spectrum (17) (see
Figure 1, upper spectrum). Permethylation g.l.c.-m.s. data
indicate this polymer to have an average repeating unit
represented by Structure 1, when n=1 -- the repeating unit then
containing three saccharide residues. Again, as for all dextrans,
the basic assumption that all (1→6)-linkages are in the backbone
is employed. The anomeric region shows three resonances of
roughly equal intensity, and progressively more linear analogues
of Structure 1 yield spectra (see Figure 4, dextran B-1402)
demonstrating that resonance 1 corresponds to the chain
extending 6-O-substituted α-D-glucopyranosyl residue. The

question now is, does resonance 2 correspond to the branching or
terminal residue?  As the terminal residue is not incorporated
into the backbone, this residue should exhibit both the general
motion of the chain, plus additional independent motion.  As the
$T_1$ value of each resonance is proportional to the relative
freedom of the carbon atom associated with that resonance, it can
be expected that the anomeric resonance associated with the
terminal residue will have the largest $T_1$ value.  Comparison of
resonance 2 vs. resonance 3 of the dextran B-1299 fraction S
spectrum clearly shows that the $T_1$ (241 millisec) is larger than
the $T_1$ (140 millisec) of peak 3.  Therefore, resonance 2
represents the anomeric resonance of the 2,6-di-O-substituted
α-D-glucopyranosyl residue (17).

A second example (17) is that of the anomeric spectral region
of dextran B-742 fraction S, a polysaccharide for which permethylation data indicate Structure 2, when n=0.  This is an
unusual polymer, as every backbone residue is 3-O-substituted.  It
is fortunate that this polymer exists, as the dextrans branching
through 3,6-di-O-substituted residues present a problem in the
anomeric spectral region, displaying only a single branching
anomeric resonance in addition to the linear dextran resonance.
It was therefore postulated that one of the two expected minor
resonances was unresolved from the linear dextran anomeric
resonance.  For dextran B-742 fraction S there are essentially
no normal linear dextran 6-O-substituted α-D-glucopyranosyl
residues, only branch-point and terminal residues -- and one of
the two approximately equal resonances appears at the same
chemical shift as the linear dextran  anomeric resonance.
Resonance number 2 of dextran B-742 fraction S has a smaller $T_1$
(150 millisec) than the corresponding resonance 1 ($T_1$= 250 millisec), and therefore resonance 2 represents the anomeric carbon
of the 3,6-di-O-substituted α-D-glucopyranosyl residue.  The
extreme structural simplicity of dextran B-742 fraction S suggests
another spectra-to-structure relationship.  The two residue
repeating unit of this dextran indicates that total resolution of
the spectrum will result in twelve (2x6) resonances and the
spectrum in Figure 6 comes close with eleven distinct resonances
resolved.  Due to a large intensity and a small $T_1$ value,
resonance 8 is believed to be a composite resonance -- leaving
10 remaining resolved resonances.  In terms of intensity, these
ten resonances break into two sets, a weak intensity set and a
strong intensity set.  The strong intensity resonance set
generally have larger $T_1$ values than the corresponding values for
the weak intensity resonance set, indication that the weak
intensity set of resonances correspond to the less mobile
backbone residue carbon atom positions.  This general assignment
of resonances is supported by the independent knowledge that
peak 3, representing a linked C-3, and peak 10, representing a
linked C-6, must be associated with the 3,6-di-O-substituted
α-D-glucopyranosyl backbone residue.  These observations then

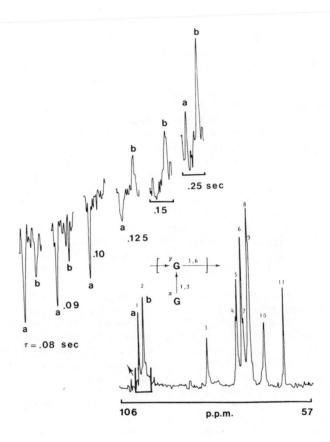

*Figure 6.   The 90° C-13 NMR spectrum of dextran B-742 Fraction S, showing the relaxation spectra for Resonance a ($T_1$ = 250 msec) and for Resonance b ($T_2$ = 150 msec): a = 101.01 ppm; b = 99.76 ppm.*

allow the introduction of another spectra-to-structure relationship, when resonances known to be associated with a specific
residue are more intense than alternative structural data would
indicate, then that residue has increased mobility and for a
polymer is probably in a terminal or side chain position.  For
example, returning to the anomeric spectral region of dextran
B-1299 fraction S, it can be noted that resonance 2 is more
intense than resonance 3.  Permethylation data indicate that all
residue types are present in approximately equal amounts, therefore resonance 2 represents the C-1 of the terminal residue (in
agreement with $T_1$ data).  Such an approach can allow the tentative
assignment of resonances when the more laborious determination of
relaxation data is not feasible.

Such $T_1$ measurements have also been employed to differentiate
between terminal and branch-point residue resonances for dextrans
branching through the 4,6-di-$\underline{O}$-substituted $\alpha$-D-glucopyranosyl
residue and for establishing that the excess 6-$\underline{O}$-substituted
$\alpha$-D-glucopyranosyl residues of the class of compounds represented
by structure 3 are in side chains relative to an alternating
linkage type backbone ($\underline{13}$).

It is known that relatively subtle solvent properties ($\underline{e.g.}$
the presence of trace metal ions or dissolved oxygen) can have
a pronounced effect on $T_1$ values ($\underline{18}$).  For this reason, we have
emphasized studies based on comparing relative $T_1$ values of
resonances taken from the same spectrum of a given compound,
rather than comparing absolute $T_1$ values taken from different
spectra.  To insure reproducibility, duplicate $T_1$ determinations
were made in all cases.  Monomer $T_1$ values (e.g. methyl $\alpha$-D-gluco-
pyranoside) can be obtained in less than an hour.  However, we
have experienced difficulty in obtaining consistent absolute $T_1$
values for successive samples of the same monosaccharide.  Such
reproducibility problems have not been observed for the polysaccharides, and we have observed no successive $T_1$ value differences which can be attributed to solvent or sample preparation.

The n.O.e. values have, in general, paralled the $T_1$ values
and provide little additional information.  Increasing n.O.e.
values reflect decreasing solvent interaction for the specific
carbon atom position associated with the resonance studied.
Interestingly, the largest n.O.e. values observed, $\sim$1.97 (near
the theoretical maximum of 1.99 for this value) is associated
with resonance 3 in the spectrum of dextran B-742 fraction S. This
resonance 3 corresponds to the linked C-3 position of the branching backbone residue, and model building, or even a casual
inspection of the drawn structure, indicates that this is an
extremely hindered position.

General conclusions

The above discussion has indicated that C-13 n.m.r. data for
a single polysaccharide can yield a large amount of structural

information.  However, the correlation of C-13 n.m.r. data from
a wide variety of polysaccharide structures can provide an even
clearer insight into spectra-to-structure relationships, and
also into the nature of the specific carbohydrate -- in fact few
of the polysaccharides we have studied have had structures so
well defined that C-13 n.m.r. could not add new information.  The
C-13 n.m.r. data have been especially welcome as they are extra-
ordinarily complementary to data obtained from permethylation
g.l.c.-m.s.   As both C-13 n.m.r. and g.l.c.-m.s. data provide
structural analysis in terms of residue types, rather than
considering an assembly of isolated linkage types, an intimate
view of the polymer structure has been obtained.

    At present, C-13 n.m.r. studies are hampered by the
relatively large sample sizes required for good signal-to-noise
ratios , and for recording the multiple spectra required to
obtain $T_1$ and n.O.e. data.  Furthermore, the extremely large
number of saccharide residues known to exist in polymers require
further development in spectral analysis of other relatively
unstudied residues -- and it is anticipated that increased
spectral resolution will be required to differentiate between
these new residue types.  However, the recent history of n.m.r.
has shown rapid advances in the introduction of spectrometers
with dramatically increased sensitivity and resolution.  There is
no reason to believe that this pace of development will slacken
in the near future.  Therefore, the general tenor of this
presentation has been to indicate that a reasonable entry into
the area of C-13 n.m.r. spectra-to-structure relationships has
been made, and that the future of these studies holds great
promise for general polysaccharide structural elucidation.

Acknowledgements

    This work was supported, in part, by a National Institutes
of Health Grant HL-17372.

Literature cited

1.  Seymour, F.R., in "Extracellular Microbial Polysaccharides",
    ACS Symposium Series No. 45, Sandford, P.A., and Laskin, A.,
    Eds., American Chemical Society, Washington, D.C., 1977,
    pp 114-127.
2.  Seymour, F.R., Plattner, R.D., and Slodki, M.E., Carbohydr.
    Res., (1975) 44, 181-198.
3.  Seymour, F.R., Slodki, M.E., Plattner, R.D., and Stodola, R.M.
    Carbohydr. Res., (1976) 48, 225-237.
4.  Seymour, F.R., Slodk-, M.E., Plattner, R.D., and Jeanes, A.,
    Carbohydr. Res., (1977) 53, 153-166.
5.  Seymour, F.R., Chen, E.C.M., and Bishop, S.H., Carbohydr.
    Res., (Unusual dextrans III), (1979) 68, 113-121.
6.  Seymour, F.R., Knapp, R.D., and Bishop, S.H., Carbohdyr. Res.,

(1976) 51, 179-194.
7.  Seymour, F.R., Knapp, R.D., Bishop, S.H., and A. Jeanes, Carbohydr. Res., (Unusual dextrans IV), 68 (1979) 23-140.
8.  Costello, A.J.R., Glonek, J., Slodki, M.E., and Seymour, F.R. Carbohydr. Res., (1975) 42, 23-37.
9.  Sidebotham, R.L., Adv. Carbohydr. Chem. Biochem., (1974) 30, 371-444.
10. Jeanes, A., "Dextran Bibliography: 1861-1976" Miscellaneous Publication No. 1355, Agriculture Research Service, U.S. Department of Agriculture, 1978, 370 pp.
11. Seymour, F.R., Knapp, R.D., Chen, E.C.M., Jeanes, A., and Bishop, S.H., Carbohydr. Res., (Unusual dextrans V) in press.
12. Jennings, H., and Smith, I.C.P., J. Am. Chem. Soc., (1973) 95, 606-608.
13. Seymour, F.R., Knapp, R.D., Chen, E.C.M., Bishop, S.H., and Jeanes, A., Carbohydr. Res., (Unusual dextrans VIII) in press.
14. Jeanes, A., and Seymour, F.R., Carbohydr. Res., (Unusual dextrans VII) in press.
15. Seymour, F.R., Knapp, R.D., Nelson, T.E., and Pfannemuller,B. Carbohydr. Res., in press.
16. Seymour, F.R., Knapp, R.D., Zweig, J., and Bishop, S.H., Carbohydr. Res., in press.
17. Seymour, F.R., Knapp, R.D., and Bishop, S.H., Carbohydr. Res., (Unusual dextrans VI) in press.
18. Stothers, J.B., "Carbon-13 Spectroscopy", Academic Press, New York, 1972, p 38.

RECEIVED March 13, 1979.

# A Reinvestigation of the Structure of Poly(dichlorophenylene oxides) Using Carbon-13 NMR Spectroscopy (1)

JOHN F. HARROD and PATRICK VAN GHELUWE

Chemistry Department, McGill University, 801 Sherbrooke St. West, Montreal, P.Q., H3A 2K6 Canada

Since their original synthesis by Hunter in 1917 (2) the poly(halophenyleneoxides) have been the subject of sporadic interest in a number of laboratories (3, 4, 5, 6, 7). In spite of their structural similarity to the useful and commercially successful poly(alkylphenyleneoxides) (8), the known halogensubstituted poly(phenyleneoxides) have poor mechanical properties and have found no commercial application. At the outset of the present study it was generally believed that the physical properties of the known polymers were impaired by a high degree of branching.

The branching hypothesis was originally based on what were thought to be anomalously low intrinsic viscosities (3) and was later reinforced by the observation of complex $^1$H-nmr spectra (4), examples of which can be seen in Figure 1. Our first suspicion that the branching hypothesis might not be correct arose from the observation that a Mark-Houwink plot of a considerable amount of viscosity/$\bar{M}_n$ data yielded parameters that were not unreasonable for a linear polymer. More importantly, some data points for polymers prepared from 4-bromo-2,6-dichlorophenoxide, known to be linear, fell on the same line as the points for supposedly branched polymers, prepared from trichlorophenoxide. (Figure 2). We therefore decided to reinvestigate the structure of poly(dichlorophenyleneoxide) using $^{13}$C-nmr spectroscopy.

Results

Molecular weight effects on $^1$H and $^{13}$C spectra. The $^1$H-nmr spectra of three samples of poly(dichlorophenyleneoxide) prepared from a copper(II)trichlorophenoxide complex and covering a molecular weight range of 5,000 to 30,000 Daltons are shown in Figure 1. By themselves it is not possible to decide whether the differences in these spectra are due to molecular weight effects or to structural differences between the various polymers. The $^{13}$C-spectra of a low and a high $\bar{M}_n$ polymer, shown in Fig.3c &d exhibit the same resonances, albeit with different linewidths and relative intensities, thereby indicating that the differences in the proton spectra are due to dynamic rather than structural effects.

Comparison of spectra of "linear" and "branched" polymers.

0-8412-0505-1/79/47-103-053$05.00/0

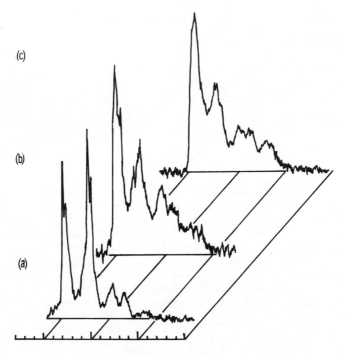

Figure 1.   Effect of $\overline{M}_n$ on ¹H-NMR spectra of polymers derived from trichlorophenoxide.  $\overline{M}_n = 6{,}000$ (a), 13,300 (b), and 33,000 (c).

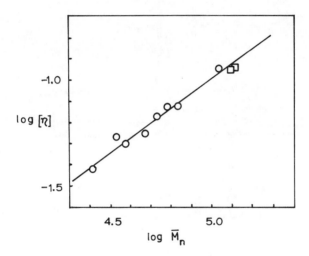

*Figure 2.   Mark–Houwink plot for polymers obtained from trichlorophenoxide*
*( ○ ) and 4-bromo-2,6-dichlorophenoxide ( □ )*

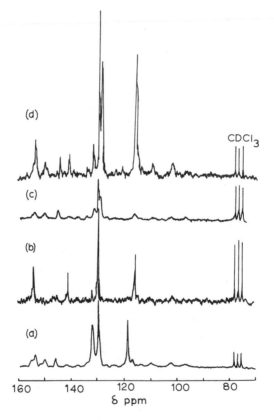

Figure 3. C-13 NMR spectra of polymers derived from (a)2-bromo-4,6-dichlorophenoxide, (b) 4-bromo-2,6-dichlorophenoxide, (c) 2,4,6-trichlorophenoxide ($\overline{M}_n \sim$ 30,000), and (d) 2,4,6-trichlorophenoxide ($\overline{M}_n = 5,000$)

As previously reported (4) linear poly(2,6-dichloro-1,4-phenyle-
neoxide) exhibits the expected sharp singlet in its [1]H-nmr spectrum, together with some minor peaks due to chemical asperities
in the polymer. The [13]C spectrum of this polymer (Figure 3) is
also very simple, consisting of the expected four sharp lines of
the four inequivalent carbons in the mer unit 1. These same four
lines may be identified

spectra of the "branched" polymers, especially that of the low molecular weight material. Subtraction of the spectrum of 1 from
that of the low $\bar{M}_n$ "branched" polymer leaves a much simpler spectrum than would be expected for an irregular, branched polymer.
In fact, six of the residual lines are exactly what one would expect for the 1,2-coupled mer unit 2. It was therefore concluded

that the so called "branched" polymer was in fact largely a copolymer of units 1 and 2 and that any unusual behaviour was a result
of the properties of unit 2 rather than branching.

    The polymer derived from 2-bromo-4,6-dichlorophenoxide and
some copolymers. Polymerization of 4-bromo-2,6-dichlorophenoxide
yields a polymer with a high proportion of units 1, indicating a
high selectivity for displacement of bromine rather than chlorine
(4,5). It was therefore anticipated that polymer derived from
2-bromo-4,6-dichlorophenoxide would contain a high proportion of
units 2. This was not the case however since elemental analysis
indicated that roughly half of the mer units in such polymers
still carried bromine. The [1]H-nmr spectra of such polymers were
indistinguishable from those of high $\bar{M}_n$ polymers derived from
2,4,6-trichlorophenoxide. The [13]C spectra were very similar
(Figure 3) except for the absence (or shifting) of a resonance at
ca. 130 ppm. and for the presence of a strong resonance at 118.6
ppm in the bromine containing polymer.

    A copolymer from equimolar 2,4,6-trichlorophenoxide and 4-bro-
mo-2,6-dichlorophenoxide with $\bar{M}_n$ > 50,000 gave both [1]H and [13]C
spectra essentially identical to those of homopolymer from 2,4,6-
trichlorophenoxide with $\bar{M}_n$ < 6000. This copolymer contained only
a trace of bromine by chemical analysis.

    A copolymer from equimolar 2-bromo-4,6-dichlorophenoxide and
4-bromo-2,6-dichlorophenoxide of $\bar{M}_n$ > 50,000 gave an [1]H spectrum

very similar to that of the homopolymer from trichlorophenoxide
with $\bar{M}_n$ < 6000 and a [13]C spectrum very similar to that of the 2-
bromophenoxide homopolymer. The copolymer spectrum was characte-
rised by sharper resonances and a considerable enhancement in intensity of the line at 116 ppm. Both chemical analysis and the
[13]C spectrum showed a level of bromine retention per 2-bromo unit
comparable to that of the 2-bromophenoxide homopolymer. The spectra of the two copolymers are shown in Figure 4.

Effects of temperature and paramagnetic additives on the [1]H
spectrum of poly(dichlorophenyleneoxide). The spectra of the polymers referred to in Figure 1 were all measured up to 120° in
tetrachloroethylene. In no case did any significant change occur
in the spectrum on heating.

The spectrum of a high molecular weight polymer was measured
in the presence of tris(acetylacetonato)chromium(III). The results for several different concentrations of reagent are shown
in Figure 5.

Discussion

The structure of poly(dichlorophenyleneoxide). Contrary to
earlier conclusions, the present evidence shows that branching is
not a major structural feature of poly(dichlorophenyleneoxide).
On the other hand it seems reasonably clear that polymer prepared
from 2,4,6-trichlorophenoxide contains substantial amounts of both
1,2- and 1,4-linked phenylene units. On the basis of this assumption all of the peaks in the [13]C spectrum can be assigned by analogy with chemically similar compounds and by the application of
additivity rules modified to take account of the shielding effects
of ethereal oxygen (9, 10, 11, 12). Such a tentative assignment
is shown in Figure 6.

Following the above conclusion it is clear that the rather
bizarre [1]H spectra of these polymers derive from the special features of 1,2-enchainment. Examination of molecular models reveals
that runs of 1,2-enchained segments are considerably more restricted in their degrees of motional freedom than are runs of 1,4- en-
chained segments. The restriction arises partly from the absence
of a "crankshaft" mode with 1,2-enchainment and partly from the
steric interference of substituents on adjacent phenylene rings.
It is probable that the 1,2- coupled units are essentially locked
in position and can only move as the whole polymer molecule moves.
The great complexity of the [1]H nmr spectra of polymers containing
1,2-coupled units could be due to the locking of chain units in a
limited number of different conformations, each with a characteristic chemical shift.

The insensitivity of the [1]H nmr spectra to large changes in
temperature is also indicative that the frozen backbone motions
responsible for the spectral complexity are not easily thermally
activated. The presence of chemically different triads composed
of 1,2 and 1,4 units, each with a particular chemical shift, could

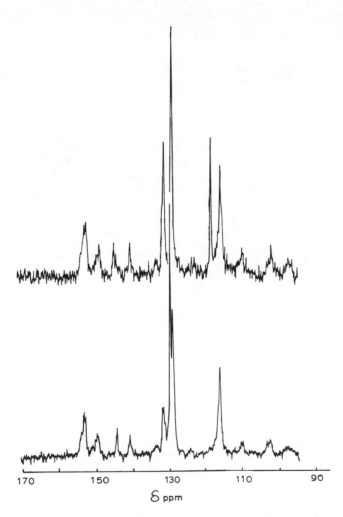

*Figure 4.  C-13 NMR spectra of copolymers derived from:* (upper) *2-bromo-4,6-dichlorophenoxide and 4-bromo-2,6-dichlorophenoxide (1:1) and* (lower) *2,4,6-Trichlorophenoxide and 4-bromo-2,6-dichlorophenoxide (1:1).*

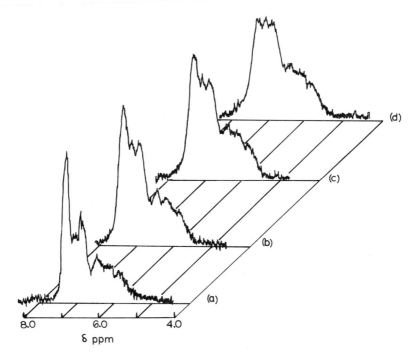

*Figure 5.* *Effect of added Cr(acac)₃ on the ¹H-NMR spectrum of polymer derived*
*from 2,4,6-trichlorophenoxide. [Cr(acac)₃] = OM (a); 5.7 × 10⁻³M (b); 1.2 ×*
*10⁻²M (c); 1.7 × 10⁻¹M (d). Polymer concentration, 17% in sym-tetrachloroethyl-*
*ene.*

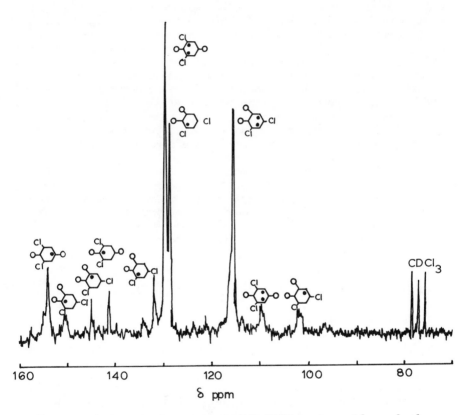

*Figure 6.   Assignment of resonances in C-13 NMR spectrum of low-molecular-weight polymer from 2,4,6-trichlorophenoxide*

also complicate the spectrum.

All of the polymers have resonances at unusually high field (>3.0 τ) suggesting that some of the protons are held in close proximity to adjacent aromatic rings. This possibility is further supported by the observation that relaxation reagents selectively collapse the lowest field peak, which is expected to be the resonance of protons least encumbered by neighboring aromatic rings and therefore more accessible to the relaxation reagent.

Unfortunately, we have found no way to estimate the relative amounts of 1,4 and 1,2-coupled units in these polymers. There are however some curious features of the results obtained with copolymers which suggest the operation of unusual selectivities in the polymerization reactions. For example, the copolymer resulting from polymerization of equimolar 4-bromo-2,6-dichlorophenoxide and 2,4,6-trichlorophenoxide ($\bar{M}_n$ > 50,000) gave both an $^1$H and a $^{13}$C spectrum very similar to that of a polymer from trichlorophenoxide with a much lower molecular weight. In the case of the $^1$H spectrum, the copolymer was quite different from a mixture of polymers derived separately from the two monomers, particularly with respect to the great reduction in the sharp singlet characteristic of polymer with a high proportion of 1,4-units (4). The $^{13}$C spectrum of the copolymer is a superposition of the spectra of the separately prepared polymers with respect to number and position of peaks, but the resonances are much sharper than those of polymer derived from 2,4,6-trichlorophenoxide and broader than those derived from 4-bromo-2,6-dichlorophenoxide (for an equivalent molecular weight). It thus appears that the deliberate introduction of more 1,4-units into the polymer chain, although leading to greater internal mobility, does not change the chemical composition in such a way as to seriously affect the chemical shifts and relative intensities of the $^{13}$C resonances. This evidence suggests that the deliberate introduction of a considerable number of 1,4 units does not significantly increase the concentration of 1,4-triads (the entity which presumably is primarily responsible for the $^1$H and $^{13}$C spectral features of the 1,4-polymer). If this interpretation is correct it means that the copolymer sequence is far from random.

The behaviour of the copolymer derived from 2-bromo- and 4-bromophenoxides was similar to that described in the previous paragraph.

Regioselectivity in the polymerization reactions. The currently accepted mechanism for the reactions used to synthesise the polymers described in the present paper involves three key steps (13):

i) a ligand transfer of phenoxyl from Cu$^{II}$ to an attacking radical.

ii) homolytic dissociation of a quinol ether.

iii) halogen atom transfer from quinol ether to Cu$^{I}$.

It is expected that all three of these steps will be sensitive to the nature of the halogen and the location of substituents in the various reacting species. Only for step (iii) can we say with any confidence what the effect will be, namely that bromine will transfer more rapidly than chlorine (14).

The high 1,4 content of polymers produced from 4-bromo-2,6-dichlorophenoxide is in accord with the expectation of easier access to the 4- relative to the 2-position in step (i) and greater reactivity of bromine in step (iii). The failure to achieve selectivity in the 2-bromo-4,6-dichlorophenoxide reaction could be attributed to the steric inaccessibility of the 2-position for step (i) overriding the greater reactivity of bromine in step (iii). This argument is not very convincing since it would lead to the conclusion that the polymer from 2,4,6-tricholorophenoxide be largely 1,4-coupled, contrary to the experimental evidence.

Chemical analysis of the polymer from 2-bromo-4,6-dichlorophenoxide revealed that somewhere between one half and two thirds of the available bromine was displaced. The presence of residual bromine in the polymer is also manifest by the sharp $^{13}C$ resonance at 118.6 ppm. This residual bromine is also the likely cause of the characteristic difference in the 130 ppm region between polymers derived from 2-bromo-4,6-dichloro (2 peaks) and 2,4,6-trichlorophenoxide (3 peaks). In addition to the units 1 and 2, the 2-bromophenoxide could also contain units 3 and 4.

Br
2  3
— O — 1   4 — O
6  5
Cl
3

Br
2  3
— O — 1   4 — Cl
6  5
— O
4

Intuitively one would expect the greater ease of displacement of bromine to strongly favour units 2 over 4 and that most of the residual bromine would be present in units 3. This could lead to the conclusion that the "missing" peak is due to the replacement of units 1 by units 3 in the 2-bromo, relative to the 2,4,6-trichlorophenoxide derived polymer. Unfortunately this hypothesis is not consistent with the spectrum of the 2-bromo/4-bromophenoxide copolymer which, despite the undoubted inclusion of many units 1, only has two peaks in the 130 ppm region.

One postulate which does resolve many of the inconsistencies in the present data base is that there is a strong tendency towards alternation in these polymers. Such alternation could result from a combination of an intrinsically greater reactivity of the ortho-relative to para-positions of coordinated phenoxide, coupled with an inability of radicals of the type 5 to attack coordinated phenoxide at the ortho-position due to steric blocking. (An examination of models shows this to be a plausible assumption).

X = Cl or Br

5                                6

In the case of trichlorophenoxide such a model would give rise to
an alternating 1,2-/1,4-sequence (which we call syndioregic by
analogy with syndiotactic). In the case of 4-bromo-2,6-dichloro-
phenoxide the much greater reactivity of the bromine relative to
chlorine overides the factor which favoured ortho over para reactivity in the trichlorophenoxide case to such an extent that 1,4
catenation results. In the case of 2-bromo-4,6-dichlorophenoxide
the 2-bromo position should be very reactive, but the radicals
produced by such reaction (type 5) are not able to attack again
at the ortho position of coordinated phenoxide.

The advantage of the alternation postulate is that it resol-
ves the anomaly that the deliberate introduction of large numbers
of units 1 into polymers derived from either 2,4,6-trichlorophe-
noxide, or from 2-bromo-4,6-dichlorophenoxide, whilst sharpening
the spectra does not radically change the characteristic numbers
and positions of the homopolymer resonances. A tendency to alter-
nate would suppress both the generation of longer runs of 1,4-cou-
pled units and the expression of the presence of such runs in the
$^1$H and $^{13}$C spectra.

A clearer understanding of the structures of these polymers
will require synthesis of model compounds for the triad and tetrad
sequences. Only with such models can more confident assignments
of the nmr spectra be made. Despite the tediousness of such syn-
theses we believe the postulate of syndioregic polymerization
to be sufficiently interesting to warrant the effort and are presently working on the problem.

Branching and the polymerization mechanism. The realisation
that the unusual properties of poly(dihalophenyleneoxides) are due
to 1,2-coupling rather than branching removes an anomaly from the
previously proposed mechanism for the polymerization reaction (13).
Following suggestions of earlier workers (3) it was assumed that
branching would occur by coupling of two polymer radicals according to equation (1). However, since the polymer radicals

1)

are rapidly scavenged by attack on coordinated phenoxide until the reactant copper phenoxide has been almost completely removed, equation (1) cannot assume any importance until late in the reaction. Since displacement of coordinated phenoxide cannot lead to branching, it would be expected that the degree of branching would increase progressively as the reaction proceeded. This was contrary to earlier experimental evidence which seemed to indicate little significant change in what was assumed to be branching, from the earliest to the latest stage of reaction.

The above anomaly is removed by the recognition that branching is not a significant factor in the reaction.

## Conclusions

The polymerization of halophenoxides by copper (II) mediated halide displacement is a mechanistically complicated reaction. Elucidation of the structure of the polymers is essential to an understanding of both the polymerization chemistry and the peculiar physical properties of the polymers. The physical tool which has yielded most information on the polymer structure is $^{13}$C nmr. The first conclusion which derives from a study of the $^{13}$C spectra of poly(dihalophenyleneoxides) is that regioselectivity in halogen displacement is more likely the source of the polymer properties than branching. A more rigorous confirmation of the polymer structures will depend on a detailed analysis of the $^{13}$C spectra of model compounds for the chain segments.

## Experimental

Synthesis of polymers. All of the polymers used in the present study were produced by methods previously described (4), or by modifications thereof. A more detailed description of the synthesis and characterisation of these polymers is in preparation.

$^1$H nmr spectra. Proton spectra were measured with a Varian T-60 spectrometer at 35°. Solutions were prepared in CS$_2$. Low molecular weight polymers were typically examined as concentrated solutions (up to 20 per cent by weight) without encountering problems with solution viscosity. High molecular weight polymers, particularly those with high proportions of 1,4-coupled units had to be run at much lower concentrations due to viscosity problems. A few test runs on a Varian HA-100 spectrometer did not yield spectra of better quality than those obtained at 60 MHz.

Variable temperature measurements were carried out on a Varian A-60 spectrometer with a Varian-4060 probe heater attachment. Probe temperatures were calibrated with a standard ethylene glycol sample. Samples were prepared by dissolving up to 25 per cent by weight of polymer in tetrachloroethylene. Even with such concentrated solutions no viscosity problem was encountered above 60°.

$^{13}$C nmr spectra. All $^{13}$C spectra were measured with a Bruker WH90 22.63 MHz pulsed FT instrument equipped with a Nicolet BNC-12 computer with 4K real data points. Probe temperature was maintained at 30°. Samples were prepared as 10 per cent solutions in deuterochloroform, which also served as the frequency lock. Chemical shifts were computed relative to an internal standard of tetramethylsilane. Spectra were proton decoupled with 5W broad band irradiation.

Acknowledgements. The authors thank the National Research Council of Canada for financial support and for a Scholarship (P.v.G.). The assistance of Dr. Gordon Hamer in the measurement of $^{13}$C spectra is gratefully acknowledged.

Literature cited.

1.  Part V in a series Chemistry of Phenoxo Complexes. Part IV, Harrod J.F. and Taylor K.R. Inorg. Chem. (1975) 14, 1541.
2.  Hunter W.H. and Joyce F.E., J. Am. Chem. Soc. (1917), 39, 2640
3.  Blanchard H.S., Finkbeiner H. and Russell G.A., J. Polymer Sci. (1962), 58, 469.
4.  Carr B.G., Harrod J.F. and van Gheluwe P., Macromolecules (1973), 6, 498.
5.  Stamatoff G.S. U.S. Patent No. 3,228,910 (Jan. 11, 1966).
6.  Tsuruya S. and Yonezawa T. J. Catalysis, (1975), 36, 48.
7.  Hedayatullah M. and Denivelle L., Compt. Rend., (1962), 254, 2369.
8.  Hay A.S., Polym. Eng. Sci. (1976), 16, 1.
9.  JEOL $^{13}$C FT-nmr spectra, Vol. 4.
10. Wilson N.K., J. Am. Chem. Soc. (1975), 97, 3573.
11. Dahmi K.S. and Stothers J.B., Can. J. Chem. (1966), 44, 2855.
12. Lauterbur P., J. Am. Chem. Soc. (1961), 83, 1846.
13. Carr B.G. and Harrod J.F., ibid, (1973), 95, 5707.
14. Kochi J., Science, (1967), 155, 415.

RECEIVED March 13, 1979.

# 4

# Carbon-13 NMR in Organic Solids: The Potential for Polymer Characterization

A. N. GARROWAY, W. B. MONIZ, and H. A. RESING

Chemistry Division, Naval Research Laboratory, Washington, DC 20375

This Symposium bears witness to the analytical power of $^{13}$C NMR as applied to organic polymers in the liquid state. But what of the solid state: can a polymer's chemistry or perhaps even engineering properties be extracted from solid state studies? A promising technique for such studies has recently evolved from the combination (1-4) of two well established NMR tools: cross-polarization (5) with dipolar decoupling (6) (proton enhanced $^{13}$C NMR (7)) and magic angle spinning (8-11). We shall outline the method and illustrate with results for epoxy polymers. Emphasis will be on NMR, for not only the potential but also the limitations of this approach should be recognized. At the outset we state that while solid state techniques produce narrow spectra and can formally imitate relaxation experiments which are well-established in the liquid state, the nature of the solid specimen may dictate (i) a fundamental restriction to spectral resolution which is independent of spectrometer limitations and (ii) a competition between spin-spin and spin-lattice processes which may hopelessly obscure the connection between the observed rotating frame relaxation rate and molecular motions. We examine first the solid state techniques, then spectral resolution and finally relaxation effects.

## MAGIC ANGLE SPINNING

In liquids rapid molecular motion serves to isolate nuclear spins from one another as the isotropic average of nuclear dipolar interactions vanishes. Also the chemical shift tensor collapses to its isotropic value. In a solid such fast isotropic motions are absent, although a partial averaging may occur through anisotropic motions, as in liquid crystals or rotating methyl groups. The averaging of these interactions provides the NMR definition of a liquid, a definition which may be at odds with the mechanical notion that a liquid cannot support a shear stress. In solids some other means must be found to reduce dipolar and chemical shift broadening (12-13).

In an organic solid representative broadenings are 150 ppm for aromatic carbon chemical shift anisotropy and 25 kHz (full width at half-height) for a rather strong carbon-proton dipolar interaction. At a carbon Larmor frequency of 15 MHz, the shift anisotropy corresponds to 2.25 kHz. In high magnetic fields the forms of the respective Hamiltonians are

$$H_{CS} = \hbar\omega_{0s} \sum_j \sigma_{zzj} S_{zj}, \tag{1}$$

$$H_D = \hbar^2 \sum_{j<k} \gamma_I \gamma_S \, I_{zj} \, S_{zk} \, (1 - 3 \cos^2 \theta_{jk}) \, r_{jk}^{-3}, \tag{2}$$

where $I_z, S_z$ are the proton and carbon spin operators along the static field direction $z$ and $r_{jk}$, $\theta_{jk}$ determine the orientation of nuclear pairs with respect to $z$. Other terms have their usual meaning. For each nucleus the chemical shift $\sigma_{zzj}$ represents the projection of the chemical shift tensor along $z$ :

$$\sigma_{zzj} = \lambda_{1j}^2 \, \sigma_{1j} + \lambda_{2j}^2 \, \sigma_{2j} + \lambda_{3j}^2 \, \sigma_{3j}, \tag{3}$$

where $\sigma_{1j} - \sigma_{3j}$ are the principal values of the shift tensor and $\lambda_{1j} - \lambda_{3j}$ are the direction cosines orienting the tensor relative to the static field. If the rigid specimen is mechanically spun at the angle $\Theta$ to $z$, then the time average of Eq. (3) is (9)

$$\bar{\sigma}_{zzj} = \frac{3}{2} \sin^2 \Theta \, \sigma_{0j} + \frac{1}{2} \, (3\cos^2 \Theta - 1) \, \sigma_{\Theta\Theta j}. \tag{4}$$

Here $\sigma_{0j}$ is the isotropic shift $(1/3 \, [\sigma_{1j} + \sigma_{2j} + \sigma_{3j}])$ and $\sigma_{\Theta\Theta}$ is the projection of the chemical shift tensor along the spinning axis and defined analogously to Eq. (3). In general this latter term produces a powder pattern when summed over nuclei at all orientations. However, when $\cos \Theta = 1/\sqrt{3}$, the powder pattern collapses and only the isotropic value remains, i.e., $\bar{\sigma}_{zzj} = \sigma_{0j}$. If this magic angle is misset by $\epsilon$ radians, then (9)

$$\bar{\sigma}_{zzj} \simeq \sigma_{0j} - \sqrt{2}\epsilon \, [\sigma_{\Theta\Theta j} - \sigma_{0j}]. \tag{5}$$

The second term leads to a vestigal powder pattern, scaled down by $\sqrt{2}\epsilon$. The rule of thumb becomes: to reduce chemical shift anisotropy by a factor of 100, set the angle to within about $\pm 1/2$ degree.

Though spectral resolution is sensitive to orientation, it is insensitive to spinning rate. Chemical shift is an inhomogeneous broadening mechanism; as the spins dephase in a coherent manner, the phase information can be recovered by a refocussing technique. The classic example of inhomogeneous broadening is the dephasing of a liquid free induction decay in an inhomogeneous static field. Here the signal can be recovered by a Carr-Purcell pulse sequence (14) with pulses applied at a rate determined by the "real" $T_2$ rather than by the field inhomogeneity. A similar criterion applies for magic angle spinning; the spinning rate need only be greater than "real" or homogeneous broadening and not necessarily greater than the chemical shift anisotropy (15). When the spinning is slow, distinct sidebands appear and in the frequency domain they trace out a tent-like pattern related to the spatial anisotropy (3,16). With increasing spinning rate, the sidebands move out and diminish in amplitude; it is convenient though not essential (16) to avoid complications of the sidebands by spinning at or above the frequency dictated by the width of the anisotropy pattern. For carbon at 15 MHz, a spinning rate of 2 kHz is adequate.

Magic angle spinning originated as a method to remove homonuclear dipolar broadening (8-11). There, however, speed *is* a necessity, and the technical problems of spinning above 10 kHz (with a centrifugal acceleration of $2 \times 10^6$ $g$'s on a 1 cm diameter sample) require great care. At 2-3 kHz, spinning is rather straight forward and conventional materials (Kel-F, Delrin, Macor and boron nitride) have been used to fabricate the rotors. Alternatively, the specimen itself can be directly machined or cast in a mold; that approach was used for the epoxies discussed here.

Two rather different spinner geometrics have evolved. The Beams (17) rotor, popularized by Andrew (9), looks like a spinning top cradled in a close fitting stator containing air jets. The Lowe and Norberg (10,11) geometry uses a right cylindrical disk supported by an axle along its center.

## DIPOLAR DECOUPLING

Proton-carbon dipolar coupling can be reduced by irradiating the protons near or on resonance with a very strong rf field. Originally called spin stirring (6), this is also known as dipolar decoupling or high power decoupling to distinguish it from the scalar decoupling of liquid state $^{13}$C NMR. If $<\omega_{CH}^2>$ is the second moment due to heteronuclear dipolar coupling, then, on irradiating protons spins $\Delta\omega_{0H}$ away from their resonance with rf field $\gamma_H B_{1H} \equiv \omega_{1H}$, the broadening collapses as (6)

$$\Delta <\omega_{CH}^2>^{1/2} = <\omega_{CH}^2>^{1/2} \cos \psi, \tag{6}$$

where $\tan^{-1} \psi \equiv \omega_{1H}/\Delta\omega_{0H}$ and $\psi = \dfrac{\pi}{2}$ is the "magic angle" for dipolar decoupling. For a small deviation $\epsilon'$ away from $\psi = \dfrac{\pi}{2}$, the residual broadening is

$$\Delta <\omega_{CH}^2>^{1/2} \simeq \epsilon' <\omega_{CH}^2>^{1/2}. \tag{7}$$

A single proton species can be irradiated exactly on resonance; in general the spread in Larmor frequencies is determined by the breadth of the chemical shift spectrum $|\Delta\sigma_H|$ $\omega_{0H}$. Chemical shift anisotropy must be included, *even* if the sample is spun. If the protons are irradiated at the center of their spectrum, then $\epsilon' \simeq \dfrac{1}{2} |\Delta\sigma_H| \omega_{1H}$. For an organic solid, take $|\Delta\sigma_H|$ as 10 ppm and a proton-carbon broadening as 25 kHz (full width at half height); then to achieve a residual width of 25 Hz at 15 MHz requires an rf field of 75 kHz. For the same residual broadening (in Hz) higher static fields mandate higher rf fields.

Even for on resonance irradiation, the coupling does not vanish identically. From a perturbation calculation, Haeberlen (18) finds the coupling is reduced by the factor

$$\Delta <\omega_{CH}^2>^{1/2} = \frac{1}{24} \frac{<\omega_{CH}^2>}{\omega_{1H}^2} <\omega_{CH}^2>^{1/2} \tag{8a}$$

$$\simeq \frac{1}{134} \left[ \frac{\Delta\omega}{\omega_{1H}} \right]^2 <\omega_{CH}^2>^{1/2} \tag{8b}$$

where Eq. (8b) is obtained by assuming the line is a Gaussian of full width $\Delta\omega$. To reduce a 25 kHz linewidth to 25 Hz requires then an rf field of 68 kHz; this correction term is independent of the size of the static field. The assumption of a Gaussian lineshape is for illustration only; the mutual proton-proton spin flips will serve to partly decouple the $I_z S_z$ interaction and make the lineshape more Lorentzian (19).

The forgoing examples indicate only two aspects of the rather complex problem of decoupling. The conclusion is that proton fields of around 75 kHz (17.5 G) are required to reduce a 25 kHz dipolar coupling by a factor of one thousand.

## CROSS-POLARIZATION

Cross-polarization is based on the notion that the vast proton spin system can be tapped to provide some carbon polarization more conveniently than by thermalization with the lattice (7). Advantages are two-fold: the carbon signal (from those $^{13}$C nuclei which are indeed in contact with protons) is enhanced and, more importantly, the experiment can be repeated at a rate determined by the proton longitudinal relaxation time $T_{1H}$, rather than by the carbon $T_{1C}$ (7). There are many variants (7) of cross-polarization and only two common ones are described below (12,20).

In *spin lock* (SL) cross-polarization the protons are first spin locked in a resonant rf field of strength $B_{1H}$. Then an rf field is applied near the carbon resonance with an amplitude selected so that

$$\gamma_C B_{1C} = \gamma_H B_{1H}. \tag{9}$$

These two fields create carbon and proton reservoirs corresponding to ordering along the respective rf fields. When the above prescription is filled (the Hartmann-Hahn condition (5)), mutual carbon-proton spin flips become energy conserving and the two reservoirs are tightly coupled with a time constant comparable to the proton-carbon $T_2$. When Eq. (9) is well met over the entire sample and for sufficiently large rf fields, the carbon signal is enhanced by a factor of four ($\gamma_H/\gamma_C$) compared to the carbon magnetization which would arise by thermalization with the lattice.

A second variation (21) initially prepares the system in a state of dipolar order; this ordering is used to cross-polarize the $^{13}$C spins. We prefer to call this *dipolar cross-relaxation* although it is also known as cross-polarization under ADRF conditions, though *a*diabatic *d*emagnetization in the *r*otating *f*rame is only one way to create dipolar order. Here the cross-polarization rate is diminished from the spin lock rate and is extremely sensitive to the proton local field. This cross-polarization rate also gives a measure of spin fluctuations which are effective at relaxing the spin lock state and hence indicates how much the measured (carbon) $T_{1\rho}^C$ arises from spin-spin rather than spin-lattice processes. This observation will be discussed further in section V.

With this background we can now illustrate and refine these notions with some experimental results for cured epoxy resins.

## SPECTRAL RESOLUTION, SPECTRAL FIDELITY AND MOLECULAR MOTION IN CURED EPOXIES

Figure 1 presents the three stages of resolution of $^{13}$C NMR for a model organic solid, the epoxy resin diglycidyl ether of bisphenol A (DGEBA) cured with piperidine. In the top spectrum, for which only a portion is shown, neither magic angle spinning nor high power decoupling were employed; the result is representative for a conventional liquid state spectrometer. In the middle figure high power decoupling is added (proton rf field of 60 kHz). Finally in the bottom trace magic angle spinning at 2.2 kHz aids the decoupling. All the chemical shift anisotropy information, previously distributed throughout the line, has now piled up at the isotropic average; not only is interpretation easier but also the signal-to-noise ratio is increased as the lines have sharpened.

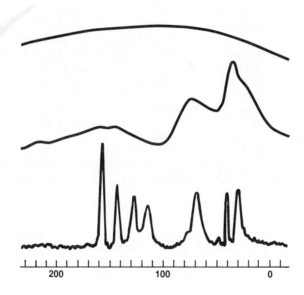

*Figure 1. Three stages of resolution in a C-13 spectrum of a cured epoxy. The top spectrum is obtained under conditions appropriate to a liquid-state spectrometer: no dipolar decoupling and no magic angle spinning. Dipolar decoupling at 60 kHz is used for the middle spectrum and to that is added magic angle rotation at 2.2 kHz for the bottom figure.*

Figure 2 indicates the potential for chemical identification of cured epoxy polymers; shown are the $^{13}$C spectra of DGEBA based epoxies reacted with four different curing agents. Commercial DGEBA (DOW DER 332) and commercial curing agents were used. Proportions and curing cycles are in Table 1. An earlier version of these spectra appears in Ref. 4.

Table 1. Composition and Cure Cycle
for DGEBA Based Epoxies.

| | |
|---|---|
| (a) Piperidine | 120°C (16 hr) |
| (PIP) 5% | |
| (b) Metaphenylene diamine | 20° (15 hr) |
| (MPDA) 13.7% | 60° (1 hr) |
| | 149° (3 hr) |
| (c) Hexahydrophthalic anhydride | 50° (16 hr) |
| (HHPA) 31.1% | 90° (2 hr) |
| N, N-dimethylbenzylamine | 120° (2 hr) |
| (DMBA) 0.2% | |
| (d) Nadic methyl anhydride | 107° (2 hr) |
| (NMA) 46.2% | 135° (2 hr) |
| N, N-dimethylbenzylamine | 166° (2 hr) |
| (DMBA) 0.8% | |

Superposed in Figure 2 are liquid state spectra (with Overhauser enhancement) and peak assignments of the *unreacted* components, in a solvent. Solvent peaks are denoted with a slash (/).

In the sample of Figure 2a the curing agent piperidine was used and the epoxy approximates polymerized DGEBA. In the solid state spectrum, peaks $f$ and $g$ of the epoxide group are not apparent, giving a crude indicator of the degree of polymerization. A previous study (22) of partially cured piperidine-DGEBA observed lines near the methylene peak ($e$) which increased in strength and complexity as the cure progressed. These peaks were attributed to the carboxyl-methine ether carbon and to the methylene carbon near the reaction site. The expanded aliphatic region in Fig. 2a shows some reaction has occured, even in the acetone solvent. (The insert spectrum was taken later than the full liquid state spectrum.) Hence in the solid all functional groups can be identified, except for the composite peak at $e$.

The contributions of the curing agents are easily seen in spectra 2b-d. Though there are unresolved peaks there is sufficient detail for chemical identification.

Two technical questions naturally arise: (i) are all the carbons counted and (ii) what limits resolution? In liquids all carbons are represented provided that the repetition period is substantially longer than the longest carbon $T_{1C}$. Resolution is generally restricted by static field inhomogeneity or by lifetime broadening. Circumstances in solids are less clearcut.

In the solid state experiments the protons are used to cross-polarize the carbon nuclei. Sufficient time must be allowed for the protons to thermalize with the lattice in between experiments but, unless the sample is inhomogeneous, all protons will share a

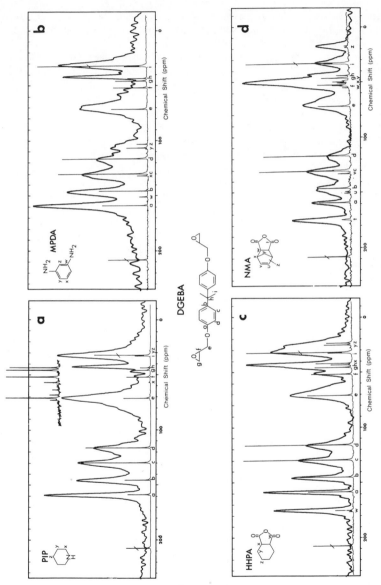

Figure 2. Solid-state spectra of four different epoxies (based on the resin diglycidyl ether of bisphenol-A) are compared with the liquid-state spectra of their respective unreacted components. The chemical compositions are in Table I. Here the epoxies are identified by their main curing agent: (a) PIP—piperidine; (b) MPDA—metaphenylene diamine; (c) HHPA—hexahydrophthalic anhydride; (d) NMA—nadic methyl anhydride.

common $T_{1H}$ and so the carbon spectrum will be undistorted, albeit slightly attenuated for repetition periods shorter than $T_{1H}$.

Let us look more closely at the cross-polarization process. It is most convenient to regard the spin locked protons and carbons as reservoirs which equilibrate by a simple first order process. This simplistic view can break down. If the protons are isolated from one another, then polarization transfer is not between two monolithic reservoirs which, by assumption, quickly reach internal equilibrium, but rather between isolated carbon-proton spin pairs. The system then oscillates coherently at a frequency determined by the CH dipolar coupling constant (5,20,23). This is the circumstance just as the two reservoirs are brought into contact; it takes approximately the proton-proton $T_2$ for spin-pairs to respond to peer pressure and reach internal equilibrium. The oscillations are most pronounced when the Hartmann-Hahn condition is severely mismatched, by an amount greater than the proton local field; the resultant of the mismatch and the dipolar spin coupling determines the frequency. In the limit that the carbons are polarized from the dipolar state, the amplitude of this oscillation is approximately $M_o < \omega^2_{CH} > / \omega^2_{1C}$ where $M_o$ is the equilibrium carbon magnetization when the two reservoirs have reached thermal equilibrium, $<\omega^2_{CH}>$ is the proton-carbon second moment, and $\omega_{1C}$ is the carbon rf field (20,21). (One difference between this and the oscillatory cross-polarization in liquids (5,24) is that the spatial dependence of the dipolar coupling renders "inequivalent" the protons in $CH_2$ or $CH_3$ groups, whereas in liquids the rotating frame cross-polarization retains the character of $AX_2$, $AX_3$ coupling (24).)

If the protons are well coupled to one another, then after a few proton-proton $T_2$'s the simple thermodynamic picture is adequate. The dynamics of equilibration are particularly pure when the protons can be considered an infinite reservoir; then the $^{13}C$ magnetization grows as (25)

$$M(t) = M_o\lambda^{-1}\{1 - \exp-(\lambda t/T_{CH})\} \exp - t/T_{1\rho}^H \tag{10a}$$

where

$$\lambda \equiv \left[1 + \frac{T_{CH}}{T_{1\rho}^C} - \frac{T_{CH}}{T_{1\rho}^H}\right] \tag{10b}$$

and where $T_{1\rho}^C$, $T_{1\rho}^H$ are the carbon and proton rotating frame relaxation times and $T_{CH}$ is the cross-polarization time under spin lock conditions. The equilibrium carbon magnetization $M_0$ is determined by the proton spin temperature. For large rf fields , $M_0 = \frac{\gamma_H}{\gamma_C} M_0^C$ where $M_0^C$ is the ordinary carbon thermal magnetization appropriate to the static field. (Here $T_{1\rho}^C$ should be defined somewhat differently: as both carbon and proton systems are irradiated, this $T_{1\rho}^C$ will be sensitive to fluctuations at $\omega_{1C} + \omega_{1H} = 2\omega_{1C}$, rather than at $\omega_{1C}$ as usual.) If the rf fields are not so large, the proton spin lock ordering is fractionally reduced (26-28) by $1/4 <\omega^2_{HH}>/\omega^2_{1H}$, where $<\omega^2_{HH}>$ is the proton second moment. This leads to a slight decline in $M_0$ but is not serious as it will influence all $^{13}C$ spins uniformly and not warp the spectrum.

Equation (10) admits a simple interpretation: the carbon magnetization rises with the rate $\lambda T_{CH}^{-1}$ while being depleted at $(T_{1\rho}^H)^{-1}$. As $T_{CH}$ is, for matched Hartmann-Hahn condition, of the order of the carbon-proton $T_2$, then $T_{CH} << T_{1\rho}^H < T_{1\rho}^C$ unless one is near a $T_{1\rho}$ minimum or unless $T_{CH}$ has been artifically increased. This behavior is illustrated in Figure 3 for the piperidine cured epoxy. The magnetizations of the chemically distinct carbons rise up with different rates and then diminish in concert. From the decay of the carbon magnetization, $T_{1\rho}^H \simeq 2.6$ ms, a value in agreement with direct observation of the proton system.

What is the influence of magic angle spinning on the cross-polarization process? In principle magic angle spinning reduces heteronuclear dipolar coupling. In order for this averaging process to be efficient, the local proton-carbon coupling must remain static for, say, one-half revolution or $(2f_{rot})^{-1}$ where $f_{rot}$ is the spinning frequency. Locally the coupling jumps around because of spin fluctuations in the proton spin system, fluctuations with a correlation time of the order of $3/<\omega_{HH}^2>^{1/2}$. Only in special circumstances (29) is $2\pi f_{rot} > <\omega_{HH}^2>^{1/2}$; for most organic solids enormous spinning rates ($\geq 10$ kHz) would be required to break completely the cross-polarization link. At more modest rates there is some effect; Figure 4 shows the cross-polarization times for protonated and non-protonated carbons in the piperidine cured epoxy for spinning speeds of 1-3 kHz. The relaxation times of the protonated aromatic, the methylene-methine and the methyl carbon increase marginally if at all. This follows from the observation that a reduction in heteronuclear line narrowing will show up only on the time scale of $(2f_{rot})^{-1}$, here 500 to 166 $\mu$s. But the cross-relaxation proceeds so rapidly ($T_{CH} \simeq 100 \mu$s for the protonated aromatics), that the cross-polarization is virtually completed by the time the spinner has rotated very far. For the unprotonated aromatic and quaternary carbons relaxation times increase by about 50% on speeding up from 1 to 3 kHz. Even so, the longest cross-polarization time (of around 500 $\mu$s at 3 kHz) is still substantially shorter than the proton $T_{1\rho}^H$ (2.6 ms). Further, at the rf field of 55 kHz, we find $T_{CH}/T_{1\rho}^C < 0.015$ for all carbons in the epoxy and so that correction in Eq. (10) can be ignored. The large gap between the $T_{CH}$'s and $T_{1\rho}^H$ insures that the $^{13}$C spectrum gives a reasonable relative indication of the carbon intensities, for contact times greater than a few $T_{CH}$'s. For more careful work, the intensities should be corrected by Eq. (10). As $T_{CH}$ approaches $T_{1\rho}^H$, the corrections become larger and, more importantly, the maximum available carbon signal is diminished by the rapid decay of the proton rotating frame magnetization.

So, for moderate spinning rates and for spin lock cross-polarization in rf fields well away from any $T_{1\rho}$ minimum, the cross-polarization spectrum counts all the $^{13}$C nuclei in contact with the proton bath. In this homily we presume that the heteronuclear coupling is not abridged by rapid, nearly isotropic motion as in a lightly cross-linked polymer above its glass transition temperature. Some estimate of the efficiency of cross-polarization can be gleaned by recognizing (5) that the Hartmann-Hahn condition transforms the heteronuclear dipolar coupling which was static in the laboratory frame into one which is static in the doubly rotating frame. That is, if a $^{13}$C resonance line (without dipolar decoupling) is not substantially broadened, then cross-polarization will not proceed. One says then that cross-polarization discriminates against liquid-like lines or indeed any carbon line not broadened by proton coupling. The discrimination of the cross-polarization process has been used to count the organic carbon rather than total carbon content of oil shale (30) and to distinguish between mobile and polymeric phases in hemoglobin (31).

Epoxies are good candidates for solid state $^{13}$C studies because of their relative chemical simplicity but even so some spectral lines overlap, as was shown in Fig. 2. We enquire into the limits of resolution to see what improvements can be expected.

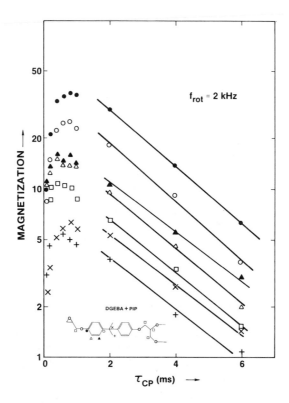

*Figure 3. Cross polarization magnetization for the PIP-cured epoxy under the SL (Hartmann–Hahn) condition. The cross polarization contact time is $\tau_{CP}$. The decay corresponds to proton $T_{1\rho}$ relaxation.*

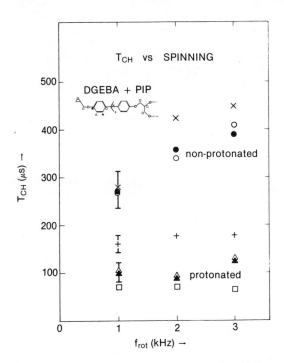

*Figure 4.   The influence of magic angle spinning on the SL cross polarization time constant. In the PIP-cured epoxy the time constant for the nonprotonated carbons increases by about 50% for $f_{rot} = 1–3$ kHz while those of the protonated carbons are virtually unchanged over this range.*

Both inhomogeneous and homogeneous line broadening can occur; in some sense the former can be refocussed by 180° rf pulses while the latter cannot. In the epoxies two prime candidates for line broadening are (i) a distribution of isotropic chemical shifts (inhomogeneous) and (ii) inadequate proton decoupling (homogeneous). Substantial fluctuation of the heteronuclear dipolar coupling, Eq. (2), from molecular motion or spin flipping at the decoupling frequency reduces decoupling efficiency.

To distinguish between these two mechanisms, create a $^{13}C$ spin echo while simultaneously decoupling, as in Fig. 5. Naively one expects any chemical shifts to refocus at $2\tau$ and hence Fourier transformation of the second half of the echo allows a value of $T_2$ (decoupled) to be determined for each line. Any diminution of the echo will then correspond to *homogeneous* broadening. (As proposed this experiment is complicated by the magic angle spinning at $f_{rot}$; the chemical shift anisotropy is reintroduced (3) to the echo when $(4\tau)^{-1} \simeq f_{rot}, 2f_{rot}$.) Alternatively, the carbon $T_{1\rho}^C$ can provide similar information; the relevant pulse sequence is also shown in Fig. 5. (As we shall no longer discuss proton relaxation, we drop the superscript and let $T_{1\rho}$ refer to carbon relaxation.)

The heteronuclear dipolar coupling between spin species $I$ and $S$ is given in Eq. (2) and rewritten below in different form:

$$H_D(t) = \sum_{j,k} I_{zj}(t) S_{zk}(t) f(\mathbf{r}_{jk}(t)), \qquad (11)$$

where $I_z(t)$, $S_z(t)$ are spin operators which may be explicitly time dependent and $f(\mathbf{r}(t))$ is a geometrical factor with argument $\mathbf{r}$ defining internuclear orientation which is implicitly time dependent. The relaxation rate in a carbon $T_2$ or $T_{1\rho}$ experiment involves the Fourier transform of the auto-correlation function of $H_D(t)$ (32). Figure 6 contrasts the two experiments. The respective rf fields impress coherent motion on the carbon spin operators $S_z$ in the $T_{1\rho}$ experiment and on the proton spin operators $I_z$ in the (decoupled) $T_2$ experiment. Molecular motions are random and so contributions to the spectral density function (the integral in the figure) will obtain for those fluctuations in $f(\mathbf{r}(t))$ at $(-\omega_{1S})$ and $(-\omega_{1I})$ respectively. When the Hartmann-Hahn condition is matched, then $T_{1\rho} = T_2$ (decoupled) in the instance that molecular motion determines the relaxation rates. (When spin-spin rather than spin-lattice effects determine the relaxation rates, then $T_{1\rho} \neq T_2$ (28).)

We then compare the measured linewidth to that implied by the "lifetime broadening" of $T_{1\rho}$ at an rf field of 66 kHz and static field of 15 MHz for the piperidine cured DGEBA epoxy at 33°C.

| | ● | O | ▲ | △ | □ | + | × |
|---|---|---|---|---|---|---|---|
| $T_{1\rho}$ (ms) | 60 | 55 | 10.5 | 9.5 | 14 | 35 | 26 |
| $(\pi T_{1\rho})^{-1}$ (Hz) | 5.3 | 5.8 | 30 | 34 | 22 | 9 | 12 |
| linewidth (Hz) | 49 | 61 | 86 | 98 | 122 | 40 | 86 |

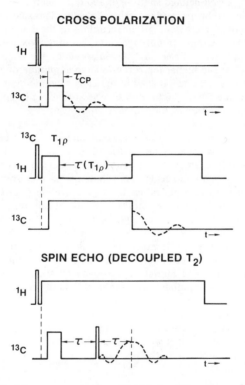

*Figure 5. The rf pulse sequences for determining SL cross polarization time constant $T_{CH}$, C-13 $T_{1\rho}$ and C-13 $T_2$ under proton decoupling. Each experiment starts with a SL cross polarization.*

At this rf field only a small fraction of the apparent linewidth can be attributed to broadening by $T_{1\rho}$ mechanism. At lower rf fields the contribution to the protonated aromatic carbon will become more important.

The source of this excess broadening must arise from chemical shift-like (inhomogeneous) terms. Magnet inhomogeneity is about 6 Hz and a missetting of the magic angle could contribute to lines with a large chemical shift anisotropy. There may also be a distribution of isotropic chemical shifts (2); evidence for this is seen on comparing the spectra for the piperidine cured epoxy at $-36$ and $+58°C$ (Fig. 7). At the lower temperature there is a well-defined splitting of about 5.5 ppm of the resonance from the protonated aromatic carbon nearest the ether linkage. The splitting likely arises from the three bond removed methylene group according to the proximity of the ortho carbon and the methylene group. In polycrystalline 1,4 dimethoxybenzene a similar splitting of the ortho carbons is observed (33), imposed by the alignment of the methyl group. In the epoxy at the higher temperature large amplitude rotation of the aromatic group averages out this splitting. Indeed in the epoxy the methyl resonance narrows by a factor of two on going from $-36$ to $+58°C$, also suggesting its chemical shift distribution is averaged by the motion of the aromatic group.

This effect — the broadening of resonance lines in highly cross-linked polymers on the order of ppm — is certainly not restricted to the epoxies. In a liquid, molecular tumbling averages out chemical shift anisotropy. But futhermore *intra* molecular motions may average *intra* molecular contributions to the isotropic chemical shift: these are not averaged to zero but rather the higher symmetry of the average chemical environment can lead to a simpler spectrum. One might call this a *conformational anisotropy*, which is unaffected by magic angle spinning. If these conformations cannot rapidly interconvert, or if they are interconverting but with a distribution of motional correlation times, then one should expect rather diffuse lines. In this event then, at a given temperature, spectral resolution will be determined primarily by the inherent chemical shift distribution and cannot be improved by spectroscopic techniques. This appears to be the case for the model epoxy at an rf field of 66 kHz.

If highest resolution is required, then the strategy is to use thermal activation to stir away the conformatonal anisotropy: operate as close to the glass transition temperature as possible, up to the point at which $T_{1\rho}$ lifetime broadening predominates. On the other hand, at lower temperatures the lineshape and its temperature dependence may provide useful information (34).

## SPIN-LATTICE RELAXATION IN THE ROTATING FRAME

Spin-lattice relaxation of $^{13}C$ nuclei is, in principle, very attractive for it is determined by local fluctuations at $\omega_{1C}$ or $\omega_{0C}$ (rotating or lab frame respectively): spin diffusion among $^{13}C$ nuclei does not average relaxation rates among chemically distinct carbons. In solids one must append a cautionary note.

Spin-spin fluctuations can compete with spin-lattice effects: an energy $\hbar\omega_{1C}$ can be supplied by a phonon as well as by a spin fluctuation in the dipolar field. A simple thermodynamic view is shown in Fig. 8. For convenience only two distinct carbon species are shown, protonated (primed) and unprotonated (unprimed). During the $T_{1\rho}$

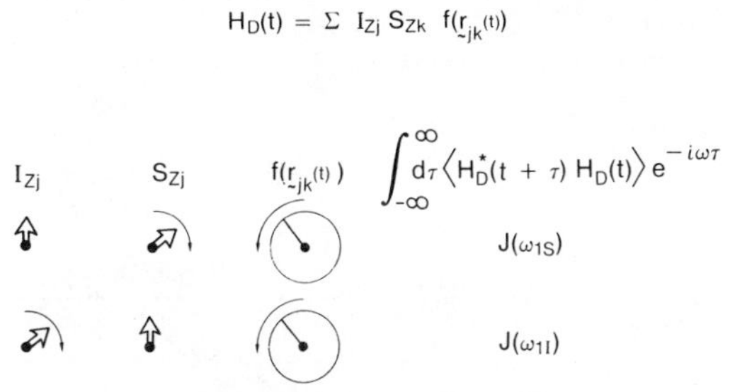

$H_D(t) = \Sigma\ I_{Zj}\ S_{Zk}\ f(\underset{\sim}{r}_{jk}(t))$
$I_{Zj}$
$S_{Zj}$
$f(\underset{\sim}{r}_{jk}(t))$
$\int_{-\infty}^{\infty} d\tau \langle H_D^*(t+\tau)\, H_D(t)\rangle e^{-i\omega\tau}$
$J(\omega_{1S})$
$J(\omega_{1I})$

*Figure 6. The proton(I)–carbon(S) dipolar coupling during a C-13* $T_{1\rho}$ *and decoupled* $T_2$ *experiment are compared. The relaxation rate is determined by the molecular fluctuation at the spin lock frequency* $\omega_{1C}$ *or decoupling frequency* $\omega_{1H}$.

200
100
0
Chemical Shift (ppm)

*Figure 7. Spectra of the PIP-cured epoxy at −36°C* (top) *and 58°C* (bottom). *The splitting of the peaks denoted by the arrow is 5.5 ppm.*

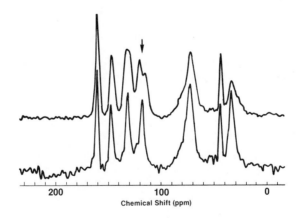

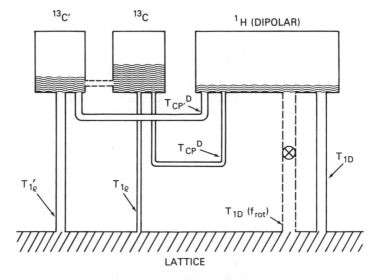

Figure 8.   During the C-13 $T_{1\rho}$ experiment, the protonated (primed) and unprotonated (unprimed) carbons are in contact with not only the lattice but also the
proton dipolar reservoir. Here $T_{1D}$ ($f_{rot}$) indicates a dipolar spin-lattice process
which depends on spinning speed and orientation.

experiment ($\omega_{1C}$ on, $\omega_{1H}$ off), the two carbon reservoirs are in thermal contact with not only the lattice by $T_{1\rho}'$ and $T_{1\rho}$ but also with the proton dipolar reservoir with $T_{CP}^{D'}$ and $T_{CP}^{D}$. The later two are cross-polarization times appropriate to magnetization transfer from the (proton) dipolar state to the carbon spin lock state (or vice versa). This pathway was first mentioned in Section III. The dipolar state is thermally linked to the lattice by spin-lattice process $T_{1D}$ and an additional process $T_{1D}(f_{rot})$ which depends on the rotational speed and orientation of the spinner (27,35,36). For an organic solid spun at around 2 kHz at the magic angle, $T_{1D}(f_{rot})$ is of the order of 100 $\mu s$ (28). The dipolar reservoir is thus strongly coupled to the lattice. For modest rf fields and for naturally abundant $^{13}C$, the specific heat of the dipolar reservoir outweighs that of any $^{13}C$ spin lock reservoir. In this limit the effective rotating frame relaxation time $T_{1\rho})_{eff}$ becomes

$$T_{1\rho}^{-1})_{eff} = (T_{1\rho})^{-1} + (T_{CP}^{D})^{-1}. \tag{12}$$

The cross-polarization (spin-spin) rate $(T_{CP}^{D})^{-1}$ competes on an equal footing with the spin-lattice (motional) rate $(T_{1\rho})^{-1}$. Now the cross-polarization rate is strongly field dependent (12,20,21,37):

$$(T_{CP}^{D})^{-1} = \frac{\pi}{2} <\omega_{CH}^{2}> \tau \exp - \omega_{1C}\tau, \tag{13}$$

where $<\omega_{CH}^{2}>$ is the proton-carbon second moment and the correlation time is determined by the proton-proton coupling (37),

$$\tau^{-2} = \frac{1}{9} <\omega_{HH}^{2}> K. \tag{14}$$

Here $<\omega_{HH}^{2}>$ is the proton-proton second moment and $K$ is a geometrical factor varying from 0.5 to 1 in crystals of cubic symmetry (37).

Rewriting Eqs. (13), (14) in terms of the proton local field $\omega_{L}^{2} \equiv 1/3 <\omega_{HH}^{2}>$,

$$T_{CP}^{D} = T_{2}' \exp \sqrt{\frac{3}{K}} \frac{\omega_{1C}}{\omega_{L}}, \tag{15a}$$

with

$$T_{2}' \equiv \frac{2}{\pi} \sqrt{\frac{K}{3}} \frac{\omega_{L}}{<\omega_{CH}^{2}>}. \tag{15b}$$

The exponential dependence holds out the prospect of finding a sufficiently large carbon rf field $\omega_{1C}$ to make $T_{CP}^{D}$ much longer than $T_{1\rho}$, which would be expected to have a $\omega_{1C}^{2}$ (38) or weaker dependence. Further away from the $T_{1\rho}$ minimum progressively larger rf fields are required to ensure $T_{CP}^{D} \gg T_{1\rho}^{C}$.

Only one complication to the determination of carbon $T_{1\rho}^{C}$ has been identified but it illustrates the role of the strongly interacting proton dipolar system, a role which must be examined in even more detail for the non-spinning case (39).

## CONCLUSIONS

The methods of high power decoupling, cross-polarization and magic angle spinning produce useful spectra from intractable polymers. In a rather rigid material increasing the decoupling power improves spectral resolution up to the limit imposed by the distribution of chemical shifts, which in a polymeric system below the glass transition temperature can be of the order of 5 ppm. At higher temperatures rapid interconversion of the conformations may reduce linewidths to a residual imposed by lifetime broadening due to the carbon $T_{1\rho}$.

In organic solids the determination of rotating frame relaxation is severely complicated by the presence of the strongly interacting proton spin system. Spin-spin fluctuations compete with spin-lattice fluctuations to produce an effective relaxation time; large rf field amplitudes are mandated to discriminate against the spin-spin event. The burden of proof lies with the experimenter to establish that a rotating frame relaxation rate actually reflects a motional effect seen by the carbon nuclei.

## Acknowledgements

Discussions with D. L. VanderHart have helped to identify the role of dipolar order in carbon rotating frame studies. This work is sponsored in part by the Naval Air Systems Command.

## Abstract

Combining magic angle spinning with proton enhanced $^{13}$C NMR, one extracts from organic solids not only $^{13}$C spectra, but also certain NMR relaxation rates. The spectra are useful for coarse chemical identification of intractable polymers although spectral resolution is limited to a few ppm in amorphous polymers; the origin of this restriction is discussed. Under certain conditions the $^{13}$C rotating frame relaxation rate $T_{1\rho}^{-1}$ as well as the spectra can reflect the nature of the molecular motions monitored at each resolvable $^{13}$C functional group. These points are illustrated in a piperidine cured epoxy based on diglycidyl ether of bisphenol A (DGEBA).

## LITERATURE CITED

(1). J. Schaefer, E. O. Stejskal, J. Am. Chem. Soc. **98**, 1031 (1976).

(2). J. Schaefer, E. O. Stejskal and R. Buchdahl, Macromolecules **10**, 384 (1977).

(3). E. Lippmaa, M. Alla and T. Tuherm, Proc. XIX Congress Ampere, Heidelberg 113 (1976).

(4). A. N. Garroway, W. B. Moniz, and H. A. Resing, Preprints of the Div. of Organic Coatings and Plastics Chem. **36**, 133 (1976).

(5). S. R. Hartmann and E. L. Hahn, Phys. Rev. **128**, 2042 (1962).

(6). (a) F. Bloch, Phys. Rev. **111**, 841 (1958), (b) L. R. Sarles and R. M. Cotts, Phys. Rev. **111**, 853 (1958).

(7). A. Pines, M. G. Gibby and J. S. Waugh, J. Chem. Phys. **59**, 569 (1973).

(8). E. R. Andrew, Arch. Sci. (Geneva) **12**, 103 (1959).

(9). E. R. Andrew, Prog. in NMR Spectroscopy **8**, 1 (1971).

(10). I. J. Lowe, Phys. Rev. Lett. **2**, 285 (1959).

(11). H. Kessemeier and R. E. Norberg, Phys. Rev. **155**, 321 (1967).

(12). M. Mehring, "High Resolution NMR Spectroscopy in Solids," NMR: Basic Principles and Progress **11** (1976).

(13). U. Haeberlen, "High Resolution NMR in Solids: Selective Averaging," Adv. in Magn. Resonance supplement **1** (1976).

(14). H. Y. Carr and E. M. Purcell, Phys. Rev. **94**, 630 (1954).

(15). E. R. Andrew, W. S. Hinshaw and R. S. Tiffen, J. Magn. Resonance **15**, 191 (1974).

(16). (a) M. M. Maricq and J. S. Waugh, Chem Phys. Lett. **47**, 327 (1977).  (b) J. S. Waugh, M. M. Maricq and R. Cantor, J. Magn. Resonance **29**, 183 (1978).

(17). J. W. Beams, Rev. Sci. Instrum. **1**, 667 (1930).

(18). Ref. 13, p. 82.

(19). (a) Ref. 12, chapt. 4.4, (b) M. Mehring, G. Sinning and A. Pines, Z. Physik B **24**, 73 (1976).

(20). D. E. Demco, J. Tegenfeldt and J. S. Waugh, Phys. Rev. B **11**, 4133 (1975).

(21). D. A. McArthur, E. L. Hahn and R. Walstedt, Phys. Rev. **188**, 609 (1969).

(22). S. A. Sojka and W. B. Moniz, J. Appl. Polym. Sci. **20**, 1977 (1976).

(23). L. Muller, A. Kumar, T. Baumann, and R. E. Ernst, Phys. Rev. Lett. **32**, 1402 (1974).

(24). R. D. Bertrand, W. B. Moniz, A. N. Garroway and G. C. Chingas, J. Am. Chem. Soc. (in press).

(25). Ref. 12, chapt. 4 discusses cross-polarization under a number of conditions; here we have included the effects of carbon $T_{1\rho}^C$ during cross-polarization.

(26). P. Mansfield and D. Ware, Phys. Rev. **168**, 318 (1968).

(27). A. N. Garroway, submitted to J. Magn. Resonance.

(28). A. N. Garroway, W. B. Moniz and H. A. Resing, to be published in Faraday Soc. Symposium **13**.

(29). E. O. Stejskal, J. Schaefer and J. S. Waugh, J. Magn. Resonance **28**, 105 (1977).

(30). H. A. Resing, A. N. Garroway, and R. N. Hazlett, Fuel, in press (1978).

(31). D. A. Torchia and D. L. VanderHart, Topics in $^{13}$C NMR **3**, in press (1979).

(32). A. Abragam, The Principles of Nuclear Magnetism, Clarendon Press, Oxford (1961), chapt. 8.

(33). E. T. Lippmaa, M. A. Alla, T. J. Pehk, G. Engelhardt, J. Am. Chem. Soc. **100**, 1929 (1978).

(34). A. N. Garroway, W. B. Moniz and H. A. Resing, in preparation.

(35). J. F. J. M. Pourquie and R. A. Wind, Phys. Lett. **55A**, 347 (1976).

(36). J. Jeener, VI International Symposium on Magnetic Resonance, Banff Canada (1977) unpublished.

(37). H. T. Stokes and D. C. Ailion, Phys. Rev. B **15**, 1271 (1977).

(38). N. Bloembergen, E. M. Purcell and R. V. Pound, Phys. Rev. **73**, 679 (1948).

(39). D. L. VanderHart and A. N. Garroway, submitted to J. Magn. Resonance.

**Discussion**

A. A. Jones, Clark University, Mass.: Could you summarize briefly those conditions under which rotating frame relaxation provides a good probe of molecular motion in the solid state? Is this something that can be achieved with a good spectrometer and a rf large rf field?

A. N. Garroway, Naval Research Laboratory, Wash., D. C.: I would like to give a simple prescription but I can't. There are a number of tests to establish that spin-lattice effects predominate: (i) direct measurement of spin-spin cross-relaxation time $T_{CP}^D$ (*cf* Fig. 8) and (ii) the sensitivity of the observed relaxation rate to rf field strength. However the spin-spin contribution must always be compared to the motional contribution and so any simple prescription for the minimum rf field required to see molecular motion necessitates *a priori* knowledge of the extent of molecular motion. As I mentioned, even with a large rf field of 80 kHz, VanderHart (Ref. 39) has not found evidence of a spin-lattice component to the carbon rotating frame relaxation in drawn polyethylene at room temperature; there is simply not very much motion in that frequency regime. Now in a system closer to the $T_{1\rho}$ minimum, the rf requirements are far less stringent.

A. A. Jones: When the specimen is spun, it is mechanically perturbed at 2-3 kHz. This is a frequency not too different from that of the motion. Does the mechanical spinning excite the system and therefore affect the rotating frame relaxation, assuming that one does everything else correctly?

A. N. Garroway: Probably not. There are two aspects. Naively what happens is that the 2 kHz spinning puts sidebands on the molecular motion. Motions which were at 60 kHz will then appear at 58 and 62 kHz. This is not of any importance provided the rf field $(\gamma B_1)$ and spinning frequency are well separated. In a system in which the dipolar strengths are of the order of only a few kHz, the effect of magic angle spinning can be seen on cross-polarization, which is a spin-spin process. A very interesting paper by Stejskal, Schaefer and Waugh (Ref. 29) addressed just this issue in adaman- tane. The second aspect is connnected with centrifugal forces. Spinning at 2 kHz does not sound ominous yet we are talking about a force of 100,000 G's on the circumfer- ence of a 1 cm diameter sample. There is some evidence at that slight changes in chemical shifts can occur. However, a few tenths of a kilobar is not really the sort of pressure one expects to change the molecular orbitals substantially. The story might well be different in an elastomer.

S. Borwnstein, NRC, Ont.: Can one use the same approach to rate processes for the motion which changes the spectra in the solid state as one normally uses in solu- tion? In Fig. 7 a doublet collapses into a singlet with temperature. This can be very easily analyzed in the liquid state to give information on the rate of a particular motion that is causing the averaging.

A. N. Garroway: Yes, however it is unlikely that a single rate is responsible.

C. J. Carmen, B. F. Goodrich, Ohio: Would you care to make any comment on the limitations of measuring carbon $T_{1\rho}$ because of the spin-spin effects? Do you feel that the practical goal understanding mechanical properties of polymer mixtures is going to be somewhat dubious as far as using carbon $T_{1\rho}$ in a real, dirty system?

A. N. Garroway: I would still hold out the hope for using these carbon relaxation rates to interpret mechanical properties, but the onus is on the experimenter to show he is actually measuring the effects of motion. Once that is done I think the idea will compete as freely as any other scientific concept.

RECEIVED March 13, 1979.

# Carbon-13 NMR Studies of Model Anionic Polymerization Systems

S. BYWATER and D. J. WORSFOLD

Division of Chemistry, National Research Council of Canada,
Ottawa, Canada K1A 0R9

The marked variation in stereostructure of diene polymers caused by changes in the counter-ion and solvent when butadiene or isoprene are polymerized anionically, are as yet not fully explained. Much progress has been made on elucidating the causes of variations in the cis/trans ratio of the 1:4 structures in these systems (1, 2, 3), but the causes of the change in the ratio of 1:2 to 1:4 structures in butadiene for example has been left largely unresolved. In dioxane, for instance, the amount of 1:2 structure decreased from 87% with Li counter-ion at 15°, to 41% with Cs (4). Less variation is found in THF because a substantial part of the reaction is carried by the free ion. Changes are also observed in polyisoprene (5).

It is likely that these changes in microstructure are a result of changes in the active chain end with the various counter ions. More specifically the decrease in the 1:2 to 1:4 ratio of polybutadiene when the series is spanned from Li to Cs might be caused by different types of bonding or changes in the charge distribution in the growing allylic carbanion. Information on bonding changes in this type of system may be sought from C-C coupling constants (6), and from the effect of deuterium substitution on the C-13 spectra. Changes of charge distribution are obtainable from variations in the C-13 chemical shifts which are sensitive to charge (7).

Hence measurements have been made on unsubstituted allyl alkali-metal compounds, and also on neopentylallyl (I, 5,5-dimethylhexen-2-) and neopentylmethallyl (II, 2,5,5-trimethyl-hexen-2-) alkali-metal compounds which are models of the polymerizing chain end in the anionic polymerization of butadiene and isoprene respectively.

## Experimental

All the allylic compounds were prepared by reacting the appropriate mercury allyl compounds with either a film of the alkali metal, or finely divided lithium, in an evacuated apparatus fitted with breakseals, a glass filter, and a nmr tube. The solvent was THF except where noted.

0-8412-0505-1/79/47-103-089$05.00/0

Allylmercury was prepared substituted in one end only of the allyl radical with either C-13 or $D_2$. The starting materials were the appropriately labelled paraformaldehydes, purchased from Merck, Sharp and Dohme. First the formaldehyde was sublimed into vinylmagnesium bromide, in THF, at $-10°C$ to give, after work up, labelled allylalcohol. The allylalcohol was not isolated but retained as a THF solution which, on distilling from $ZnCl_2$ and conc. HCl, gave a THF solution of allylchloride. At this point the labelling was scrambled between the two end positions. After drying on $CaH_2$, the allylchloride solution was used to prepare allylmagnesium chloride which on reacting with mercuric chloride gave diallylmercury. The product was distilled at low pressure, to give an overall yield of 14%.

The mercury compounds corresponding to I and II were prepared as described before (3).

All nmr measurements were made on a Varian XL100 instrument. Shifts are reported in ppm downfield of TMS.

Results

The allyl alkali-metal compounds give 2 line C-13 nmr spectra. The terminal positions are equivalent either because of rapid equilibrium between two covalent structures, or because the structure is a delocalized symmetrical ion. The chemical shifts and C-C coupling constants are recorded in table I. Reasonable agreement with literature δ values are found (8,9).

Deuteration of one end of the allyl moiety in these compounds removes the equivalence of the two positions and in place of the single line for the terminal position, two separate absorptions should appear (10). One, in the normal decoupled spectrum, is a singlet for the hydrogen substituted carbon, and the other a weak quintet for the deuterated end which would be difficult to observe. These two signals would bracket the normal singlet. If a mixture of deuterated and undeuterated allyl compound is used, therefore, two easily observable peaks should appear, one in the normal position, the other shifted. In the spectra of allyllithium and allylsodium the line from the deuterated compound appeared 14 and 11 Hz upfield respectively, at 0°C, of the normal lines. The potassium compound only showed a somewhat broadened line. At -80°C the separation for allyllithium was 22 Hz.

C-13 nmr spectra were taken of all the alkali metal compounds of I and II from Li to Cs. As expected, changing the counter-ion had very little effect on the chemical shifts of the carbons in the neopentyl group in either I or II, or on the extra methyl group in II, compared with the parent hydrocarbon (3). The substantial variation in the shifts of the allyl carbons are shown in table II. In both series the β carbon is moved 10-14 ppm downfield from its position for the parent hydrocarbon. The position of the γ peak moves markedly upfield from Li to K and then remains fairly constant, while the α carbon moves downfield over the whole series.

## Table I.

Allyl M [13]C Chemical Shifts[a] and Coupling Constants[b] at 0°C

| M | $\delta C_1$ | $\delta C_2$ | $J_{C-C}$ |
|---|---|---|---|
| Li | 50.8 | 146.0 | 55.9 |
| K | 52.4 | 143.3 | 59.8 |
| Cs | 60.0 | 143.6 | 61.1 |

a.  In THF solution, the shifts recorded are as ppm downfield of TMS, using THF $\delta$ = 26.21 as standard

b.  in hertz

## Discussion

If the allyl alkali metal compounds are ionic compounds, then if the allyl ion is delocalized the terminal carbons would be equivalent and contain most of the charge, and all three carbons would be $sp^2$ hybridized.

[13]C-[13]C nmr coupling constants have been shown to be primarily sensitive to the s character of the bonded carbons. In simple aliphatic compounds most $J_{C-C}$ values of $sp^3$-$sp^2$ bonded carbons lie between 40 to 60 hz, whilst those between two $sp^2$ hybridized carbons are in general above 65 hz. Aromatic compounds, however, such as benzene and its derivatives seem to have intermediate values of 55-60 hz for the aromatic carbon coupling constants.

Studies on benzyl alkali metal compounds, models of styrene anionic polymerization systems, showed coupling constants to the enriched α carbon in the above ranges, showing it to be $sp^2$ hybridized (11). Little effect of the charge could be detected in the $J_{C-C}$ values, although it should be noted that the charge is extensively delocalized into the benzene ring.

In the present study, the $J_{C-C}$ values in the allyl alkali metal compounds are near 60 hz for the K and Cs compounds, and a little lower for the Li compound, table I. Allyl mercury is not a delocalized system, and if any exchange occurs between the two ends of the allyl radical it is slow on the NMR time scale. Consequently $J_{\alpha-\beta}$ and $J_{\beta-\gamma}$ are measurable and are 42 and 69 hz respectively. Thus taking the mercury compound as a model of a static covalent structure, rapidly equilibrating allylic compounds would be expected to give an averaged coupling constant near 55 hz which is close to the Li compound's value, and not very far from those of the K and Cs compounds. This predicted value is not sufficiently different from that observed in delocalized structures such as benzene to distinguish between these two

possible types of structure although higher coupling constants
are observed in other olefinic cpds (∿70 cps). But the evidence
of West(12) on non-equivalence of the hydrogens at -80° in allyllithium and the small probability of allyl cesium being covalent
suggest that for allylic $sp^2$ hybridized species coupling
constants lower than 70 cps are to be expected. Even lower
values were observed for compound II-Li in THF (46 hz) and in
benzene (36 hz) (11). The latter value is even below the
reported range for systems having the $sp^3$-$sp^2$ hybridization
expected of covalent compounds. Uncertainty as to the effect of
association and charge on $J_{C-C}$ make it unwise to speculate at
this time on the meaning of these results.

It has been suggested that the effect of selective
deuteration on C-13 spectra of symmetrical carbenium ions can be
used to distinguish between rapidly equilibrating structures
(ie. involving hydride and methide shifts) and delocalized
structures (13).

Similar effects could be expected for carbanions. Normally
$k_H/k_D$ ratios in the region 1.15-1.20 are found per deuterium in
the formation of carbenium ions. If this ratio is translated
into the separation of C-13 shifts expected by deuteration of one
side of an equilibrating ion, then $\delta/\Delta$ ∿ 0.1 per deuterium, where
$\delta$ is the difference between the observed lines and $\Delta$ is the
separation expected between the shifts of the two positions in
each tautomeric structure (10). If, however, there is resonance
between the two structures, this ratio appears to be considerable
smaller (14).

This ratio calculated from the observations on the
deuterated allyl alkalimetal compounds is 0.0044 for Li, 0.0034
and perhaps half this for K. This is sufficiently smaller than
the 0.1 figure quoted above to give support to the delocalized
form of all these allyl alkalimetal compounds including the
lithium compound.

The upfield movement of the C-13 chemical shifts (table II)
of compounds I and II, models of the polymerizing systems,
indicates that the negative charge residing of the γ carbon
increases from Li to K, and then remains fairly steady for Rb
and Cs. The movement of the chemical shifts of the α carbon is
in the opposite direction, which would indicate a lessening of
charge at this position. As the movements are roughly comparable
in magnitude this suggests a transfer of charge from the α to the
γ position as the counter ion increases in size. $Li^+$ is small
relative to the allylic system, but $Cs^+$ can overlap all three
positions. Thus although the $Li^+$ ion could well localize the
charge at a particular position (α), the $Cs^+$ ion could allow a
charge distribution more nearly approaching that of the free
anion.

Although all change of the γ carbon chemical shift from that
of the parent hydrocarbon in compounds I and II is caused by
charge, the overall change in the α position's shift is composed

Table II.

$^{13}$C Chemical Shifts[a] of Allyl Carbons in I-M[b] and II-M[b]

| Cpd. | Solvent | $\delta_\alpha$ | $\delta_\beta$ | $\delta_\gamma$ |
|------|---------|------|------|------|
| I-H | THF | 12.8 | 125.3 | 127.8 |
| I-Li | $C_6D_6$ | 20.0 | 140.3 | 103.0 |
| I-Li | DEE | 30.7 | 140.0 | 87.6 |
| I-Li | THF | 31.0 | 142.5 | 81.9 |
| I-Na | THF | 35.7 | 138.8 | 72.3 |
| I-K | THF | 45.0 | 137.5 | 67.5 |
| I-Rb | THF | 47.4 | 138.2 | 67.5 |
| I-Cs | THF | 51.4 | 139.5 | 69.0 |
| II-H | $C_6D_6$ | 18.0 | 132.4 | 122.3 |
| II-Li | $C_6D_6$ | 24.2 | 148.4 | 100.4 |
| II-Li | DEE | 33.3 | 149.2 | 87.6 |
| II-Li | THF | 31.8 | 149.3 | 83.9 |
| II-Na | THF | 36.7 | 146.1 | 79.2 |
| II-K | THF | 45.8 | 143.2 | 71.9 |
| II-Rb | THF | 49.0 | 143.1 | 69.0 |
| II-Cs | THF | 53.7 | 143.4 | 70.8 |

a. Shifts are recorded in ppm downfield of TMS using the following solvent peaks as standards. $C_6D_6$ = 128.0; Diethylether (DEE) = 15.55, THF = 26.10. Temperature -20° except for parent hydrocarbons and $C_6D_6$ solutions which were measured at +20°.

b. Chemicals shifts are quoted only for the cis isomer. In many cases the trans isomer is not observed.

of two contributions. There is an upfield component due to
negative charge, and a downfield movement caused if there is a
change in hybridization from $sp^3$ in the parent compound to $sp^2$
in the ion. The magnitude of these two effects can be obtained
by using allylpotassium as a model of a delocalized ionic
compound and comparing its chemical shifts with those of propene.
The upfield movement of the $\gamma$ shift is 63 ppm, whilst the $\alpha$ shift
is 34 ppm downfield. Hence in this symmetrical ion this 34 ppm
must be composed of the 63 ppm upfield movement (as in the $\gamma$
position) caused by charge, and an extra 97 ppm downfield movement
caused by the hybridization change. The figure 97 ppm is a
reasonable figure for such a $sp^3 \rightarrow sp^2$ change ($\delta C_1$, in propane and
propene is 15.6 and 115.0 respectively). As the $\beta$ carbon has a
change of chemical shift of 11 ppm downfield, the net upfield
movement of chemical shifts caused by charge is 115 ppm. If this
is equated to $1\varepsilon$, then the upfield shift per electron of 115 ppm
is substantially smaller than that found for aromatic systems. A
similar calculation on allyllithium gives a figure of 114 ppm per
electron.

Using these figures, it is then possible to calculate the
charges on the alkalimetal compounds I and II, and these are
shown in Table III. It is seen that the total charge approaches
$1\varepsilon$ as expected, at least for the higher alkalimetal compounds.
The charge distribution does change markedly. There is a
localization of the charge at the $\alpha$ position in the Li compounds,
but for Cs there is almost equal distribution of charge between
the two positions. Despite the evidence of low coupling
constants, the Li compounds appear to be in reasonable agreement
with this scheme even in benzene solution.

Nmr measurements in THF are characteristic of the ion-pairs
(presumably contact). Microstructure measurements in this
solvent are affected by the very small amount of very reactive
free anions present in dilute solutions. These produce with
isoprene (15) (and probably butadiene) very high vinyl contents
in the polymer. The vinyl content in THF (16) does not drop for
this reason as rapidly with increasing counter-ion size as in
diethylether and dioxane (Table IV). Nevertheless it can be
qualitatively assumed (as did Essel (5)) that each ion-pair
produces a characteristic microstructure largely independent of
solvent. Such a generalization would be expected to be least
valid for lithium where changes in external solvation could
produce some differences. It appears generally therefore that in
ether solvents there is an inverse relationship between vinyl
content and the charge on the $\gamma$-position of the active centre
(the reaction site for this structure). The $\gamma$-position is
however much more reactive as indicated by the free anion product.
The charge distribution in this case must surely be very similar
to that observed with $Cs^+$ as counter-ion, yet the polymers
produced are quite dissimilar. Steric accessibility of the
$\gamma$-position must for the larger counter-ions be a major directing

Table III.

Calculated Charges[a] on Allylic Positions

| M | Solvent | COMPOUND I-M | | | | COMPOUND II-M | | | |
|---|---------|------|------|------|------|------|------|------|------|
| | | $\alpha$ | $\beta$ | $\gamma$ | $\Sigma$ Total | $\alpha$ | $\beta$ | $\gamma$ | $\Sigma$ Total |
| Li | $C_6H_6$ | .79 | -.13 | .22 | .88 | .80 | -.14 | .19 | .85 |
| | DEE | .69 | -.13 | .35 | .91 | .72 | -.15 | .30 | .87 |
| | THF | .69 | -.15 | .40 | .94 | .73 | -.15 | .34 | .92 |
| | DME | | | | | .72 | -.14 | .35 | .93 |
| Na | THF | .65 | -.12 | .49 | 1.02 | .69 | -.12 | .38 | .95 |
| K | THF | .59 | -.11 | .53 | 1.01 | .61 | -.09 | .44 | .96 |
| Rb | THF | .55 | -.11 | .53 | .97 | .58 | -.09 | .47 | .96 |
| Cs | THF | .51 | -.12 | .52 | .91 | .54 | -.10 | .45 | .89 |

[a]Charges tabulated as fractions of 1 electron at that position.

Table IV.

Effect of Counter Ion on the % 1,2 Structure in Polybutadiene

| Counter ion | Solvent | | |
|---|---|---|---|
| | DEE | Dioxan (4) | THF (15) |
| Li | 73 | 87 | 96 |
| Na | | 85 | 91 |
| K | 58 | 55 | 83 |
| Rb | | | 75 |
| Cs | 44 | 41 | 74 |

influence in the reaction. The ability of the lithium ion to reside close to the $\alpha$-position must leave the $\gamma$ more open to reaction. The large $Cs^+$ can overlap both positions which apparently causes increased attack at the $\alpha$ position, perhaps by facilitating an end approach.

Literature Cited

1.  Garton, A. and Bywater, S., Macromolecules, 8, 697 (1975).
2.  Gebert, W., Hinz, J. and Sinn, H., Makromol. Chem., 144, 97 (1971).
3.  Bywater, S. and Worsfold, D.J., Macromolecules, in press.
4.  Salle, R. and Pham, Q-T., J. Polym. Sci., Chem. Ed., 15, 1799 (1977).
5.  Essel, A. and Pham, Q-T., J. Polym. Sci., A-1, 10, 2793 (1972).
6.  Frei, K. and Bernstein, H.J., J. Chem. Phys., 38, 1216 (1963).
7.  Spiesecke, H. and Schneider, W.G., Tetrahedron Lett., 468 (1961).
8.  van Dongen, J.P.C.M., van Dijkman, H.W.D. and Bie, M.J.A., Recl. Trav. Chim. Pays Bas, 29, 93 (1974).
9.  O'Brien, D.H., Russell, C.R. and Hart, A.J., Tetrahedron Lett., 37 (1976).
10. Saunders, M., Telkowski, L. and Kates, M.R., J. Am. Chem. Soc., 99, 8070 (1977).
11. Bywater, S., Patmore, D.J. and Worsfold, D.J., J. Organometal. Chem., 135, 145 (1977).
12. West, P., Purmort, J.I. and McKinley, S.V., J. Am. Chem. Soc., 90, 797 (1968).
13. Saunders, M., Kates, M.R., Wiberg, K.B. and Pratt, W., J. Am. Chem. Soc., 99, 8072 (1977).
14. Saunders, M. and Kates, M.R., J. Am. Chem. Soc., 99, 8071 (1977).
15. Bywater, S. and Worsfold, D.J., Can. J. Chem., 45, 1821 (1967).
16. Rembaum, A., Ells, F.R., Morrow, R.C. and Tobolsky, A.V., J. Pol. Sci., 61, 166 (1962).

Discussion

J. Prud'homme, U. of Montreal, Quebec: With THF as solvent we see that the 1:2 structure is about constant, it does not change much. What would you expect for dioxane solutions?

D.J. Worsfold: In dioxane solution the polymer produced is from reaction of the ion pair almost entirely. As the NMR spectra are taken in concentrated solution, even in THF the structure observed is the ion pair and should correlate with the polymer structure formed in dioxane. In polymerizations in THF solution the polymer is quite largely formed from the reaction of the free ion chain end, and hence is less dependent on the ion pair structure. The NMR spectra were measured at -20°C because these compounds are not very stable, but as dioxane freezes at +10°C it was not possible to use it for the NMR solvent.

RECEIVED March 13, 1979.

# Carbon-13 NMR High-Resolution Characterization of Elastomer Systems

CHARLES J. CARMAN

The B. F. Goodrich Research & Development Center, Brecksville, OH 44141

Identification and characterization of elastomers are important analytical information for those involved in rubber technol-
ogy. It has been shown (1) that Fourier transform carbon-13
nuclear magnetic resonance ($^{13}C$ nmr) spectroscopy is a very
powerful method for determining subtle molecular structure features in elastomers. Details from typical information-rich
spectra can be used to describe monomer composition and sequence
distribution, chain configuration or simply materials identification of elastomer mixtures. These types of analyses are usually
made from high resolution solution spectra. That is, a rubber is
dissolved in a solvent with noninterfering resonances and the
spectrum obtained at high temperature. This produces a spectrum
typically with linewidths three to five hertz wide, and consequently has maximum chemical shift information. The advantage of
obtaining $^{13}C$ nmr spectra on solid elastomers was suggested by the
early report (2) of a high quality spectrum of solid natural
rubber gum, and of the high resolution spectrum obtained when the
$^{13}C$ nmr analysis was combined with magic angle spinning (3) on a
cured carbon black filled natural rubber sample. This paper will
compare some of the structural information that one can obtain
from a solution spectrum with the information available from
spectra obtained directly from solid elastomers. Spectra obtained
from normal Bloch decays using the conventional proton decoupling
typical for organic compounds will be compared to spectra obtained
with magic angle spinning and high power proton decoupling. It
will be shown that a combination of solution spectra of raw elastomers with solid spectra of cured products usually can be used
for materials identification.

## Results and Discussion

High Resolution Spectra of Solutions. An example of high
resolution solution spectra of an elastomer system which illustrates the sensitivity of $^{13}C$ nmr to molecular structure is shown
in Figure 1. Shown are spectra of ethylene propylene rubbers

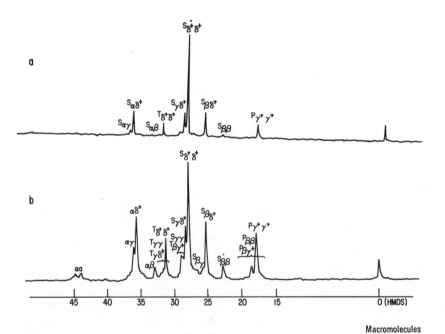

*Figure 1.   Pulsed FT C-13 NMR spectrum of (a) a 71 wt. % $C_2H_4$ and (b) a 56 wt % $C_2H_4$ EPDM rubbers obtained at 393 K from a trichlorobenzene solution. (Figure 1b reproduced from Ref. 11.)*

having different monomer compositions.

Chemical shift assignments in these complex spectra were made (4,5,6,7,8) using the Grant and Paul (9) relationships and alkane model compounds (10). Separate $^{13}C$ resonances are obtained for the methyl, methine and methylene carbons. The fact there are many resonances is because each methylene sequence, as defined by the tertiary carbon branch point, produces a unique resonance for each carbon in the sequence. We developed a mathematical model (11), based on reaction probability, which accurately accounts for all resonances in a spectrum of an ethylene propylene rubber.

The reaction probability model produces a complete calculated $^{13}C$ nmr spectrum as a best fit to the observed experimental spectrum. The deduced probabilities provide the following derived quantities: (1) "$r_1r_2$", a measure of monomer sequence randomness, (2) the distribution of methylene sequence lengths, (3) composition, (4) amount of propylene inversion.

We considered the polymer chain could be formed by either primary or secondary insertion. Primary insertion is when the monomer forms a bond with the methylene group to the catalyst metal and secondary insertion is where the metal and monomer form a bond with the propylene methine group. Ethylene must always add by primary insertion, but propylene can add either way. And, in fact, the presence of the $\alpha\beta$ secondary carbon shows the presence of a two carbon methylene sequence which can occur only if both types of insertion are in force.

Table I shows the $^{13}C$ chemical shifts and their assignments in terms of secondary, primary and tertiary carbons. Also shown is the sequence with which each species is associated. We express each sequence of methylene for all possible chain lengths in terms of conditional probabilities. As an example, consider a sequence of length three. It can be formed in two ways:

$$\rfloor \_ \rfloor \text{ and } \lfloor \_ \lfloor$$

The number of the first kind is the product of type 2 propylene ($N_2$) and the probability ($p_{21}p_{12}$). Similarly, the number of the second kind is the product, $N_3p_{31}p_{13}$. The result is that sequences of three methylene carbons can be described by

$$s_3 = N_2p_{21}p_{12} + N_3p_{31}p_{13}$$

Table II gives our description for each occurring methylene sequence in terms of conditional probabilities. We take a measured $^{13}C$ nmr spectrum of an ethylene propylene elastomer and derive a set of six reaction probabilities that fit it best. For possible sets of probability values, a spectrum is calculated and a sum of weighted squares of errors is found. The set of probability values having the smallest sum of weighted squares of error is used as a starting point to minimize the error using the

## TABLE I.

$^{13}$C Chemical Shifts for Ethylene Propylene Rubbers

| Species | $^{13}$C NMR Shift[a] | Occurrence |
|---|---|---|
| Sαα | 44.6-43.7 | methylene sequence length 1 |
| Sαβ | 32.9 | two in each sequence length 2 |
| Sαγ | b, 35.9, 36.4 | two in each sequence length 3 |
| Sαδ⁺ | 35.5 | two in each sequence length M>3 |
| Sββ | 22.7 | methylene sequence length 3 |
| Sβγ | 25.8 | two in each sequence length 4 |
| Sβδ⁺ | 25.4 | two in each sequence length M>4 |
| Sγγ | 28.8 | methylene sequence length 5 |
| Sγδ⁺ | 28.4 | two in each sequence length M>5 |
| Sδ⁺δ⁺ | 28.0 | M-6 in each sequence length M>6 |
| Tαβ⁺ | c (38.9) | two for each ⌐⌐ |
| Tββ | 27.0 | ⌐⌐⌐ and ⌐⌐⌐ |
| Tβγ⁺ | b, 29.0, 28.8 | _⌐⌐,_⌐⌐,⌐⌐⌐_,⌐⌐⌐,⌐⌐⌐,⌐⌐⌐_ |
| Tγγ | 31.6 | ⌐⌐_⌐ and ⌐_⌐⌐ |
| Tγδ⁺ | 31.6 | __⌐⌐,_⌐_⌐,⌐_⌐⌐,⌐⌐__,⌐⌐_⌐, ⌐_⌐_ |
| Tδ⁺δ⁺ | 31.3 | __⌐_,_⌐__,_⌐_⌐,⌐_⌐_ |
| Pαβ⁺ | c (22.9) | attached to Tαβ⁺ (⌐⌐) |
| Pββ | b, 19.8, 19.6, 18.9, 18.7 | attached to Tββ (⌐⌐⌐ and ⌐⌐⌐) |
| Pβγ⁺ | 18.7 | attached to Tβγ⁺ (_⌐⌐,⌐⌐_ etc.) |
| Pγ⁺γ⁺ | 18.2 | attached to Tγγ, Tδδ⁺, and Tδ⁺δ⁺ |

[a] downfield from internal HMDS (hexamethyldisiloxane) at 120°C in OCDB or TCB

[b] stereostructure produces non-equivalent chemical shifts

[c] not detected; predicted (10) chemical shift is shown in parentheses

(reproduced from C. J. Carman, R. A. Harrington, C. E. Wilkes, Macromolecules, (1977), 10, 536.)

## TABLE II.

### Numbers of Methylene Sequences of Different Lengths

| Length | Occurrence | Number in "representative sample" |
|--------|-----------|-----------------------------------|
| 1 | └ └ and ┘ ┘ | $S_1 = N_2 P_{22} + N_3 P_{33}$ |
| 2 | └ ┘ and ┘ └ | $S_2 = N_2 P_{23} + N_3 P_{31} P_{12}$ |
| 3 | └ _ └ and ┘ _ ┘ | $S_3 = N_2 P_{21} P_{12} + N_3 P_{31} P_{13}$ |
| 4 | └ _ ┘ and ┘ _ _ └ | $S_4 = N_2 P_{21} P_{13} + N_3 P_{31} P_{11} P_{12}$ |
| 5 | └ _ _ └, ┘ _ _ ┘ | $S_5 = P_{11} S_3$ |
| 6 | └ _ _ _ ┘, ┘ _ _ _ └ | $S_6 = P_{11} S_4$ |
| 3+2n | └ └ and ┘ ┘ | $S_3 + 2n = P_{11}{}^n S_3$ |
| 4+2n | └ ┘ and ┘ └ | $S_4 + 2n = P_{11}{}^n S_4$ |
| all | | $\Sigma s = N_2 + N_3 - N_3 P_{32}$ |

(reproduced from C. J. Carman, R. A. Harrington, C. E. Wilkes, Macromolecules, (1977), 10, 536.)

method of steepest descent. Table IV shows how well the peak areas of a $^{13}C$ nmr spectrum of ethylene propylene rubber (EPDM A) can be fitted using the conditional probability model. The utility of the method is that rubbers of different composition can be analyzed and compared in terms of their composition as well as in terms of relative monomer sequence distribution.

Table V gives a comparison of the two rubbers whose spectra were shown in Figure 1. The compositions are a result of the analysis and are comparable with those obtained from proton nmr or infrared (1). The information on sequence structure, however, is unique to the $^{13}C$ nmr analysis and is testimony to its power to obtain molecular structure information. The value for "$r_1 r_2$" shows that these rubbers are not perfectly random but tend toward altenation. The amount of propylene inversion could not be quantitatively measured before the advent of $^{13}C$ nmr even though infrared could be used to estimate its presence (12). One can see that the amount of propylene inversion is not constant and $^{13}C$ nmr analysis (11,13) shows this can run quite high.

Table VI yields the methylene sequence distribution of these two EPDM rubbers. A measure of the fraction of long methylene runs is important to the understanding of physical properties of these and similar materials. Polyethylene blends with EPDM rubbers, having appropriate long runs of ethylene, have been shown (14,15) to have unusually high tensile strengths.

## TABLE III.

Number of Each Carbon Species as a Function of Sequence

| Code | Species | Number |
|------|---------|--------|
| A | $S\alpha\alpha$ | $s_1$ |
| B | $S\alpha\beta$ | $2s_2$ |
| D | $S\alpha\gamma$ | $2s_3$ |
| E | $S\alpha\delta^+$ | $2(\Sigma s - s_1 - s_2 - s_3)$ |
| C | $S\beta\beta$ | $s_3$ |
| F | $S\beta\gamma$ | $2s_4$ |
| G | $S\beta\delta^+$ | $2(\Sigma s - s_1 - s_2 - s_3 - s_4)$ |
| H | $S\gamma\gamma$ | $s_5$ |
| I | $S\gamma\delta^+$ | $2(\Sigma s - s_1 - s_2 - s_3 - s_4 - s_5)$ |
| J | $S\delta^+\delta^+$ | $2N_1 + N_2 + N_3 -$ (sum of above numbers) |
| - | $T\alpha\beta^+$ | $2N_3 p_{32}$ |
| K | $T\beta\beta$ | $N_2 p_{22}^2 + N_3 p_{33}^2$ |
| L | $T\beta\gamma^+$ | $N_1 p_{12} p_{22} + N_1 p_{13} p_{33} + N_2 p_{22}(p_{21} + p_{23}) +$ $N_2 p_{23} p_{33} + N_3 p_{33} p_{31}$ |
| M | $T\gamma\gamma$ | $N_2 p_{23} p_{31} p_{12} + N_3 p_{31} p_{12} p_{23}$ |
| N | $T\gamma\delta^+$ | $N_1 p_{11} p_{12} p_{23} + N_1 p_{13} p_{31} p_{12} + N_2 p_{21} p_{12} p_{23} +$ $N_2 p_{23} p_{31}(p_{11} + p_{13}) + N_3 p_{31} p_{12} p_{21}$ |
| O | $T\delta^+\delta^+$ | $N_1 p_{11} p_{12} p_{21} + N_1 p_{13} p_{31}(p_{11} + p_{13}) +$ $N_2 p_{21} p_{12} p_{21}$ |
| - | $P\alpha\beta^+$ | same as $T\alpha\beta^+$ |
| P | $P\beta\beta$ | same as $T\beta\beta$ |
| Q | $P\beta\gamma^+$ | same as $T\beta\gamma^+$ |
| R | $P\gamma^+\gamma^+$ | sum of numbers for $T\gamma\gamma$, $T\gamma\delta^+$, and $T\delta^+\delta^+$ |
| | all S | $2N_1 + N_2 + N_3$ |
| | all T | $N_2 + N_3$ |
| | all P | $N_2 + N_3$ |
| | all species | $2N_1 + 3N_2 + 3N_3$ |

(reproduced from C. J. Carman, R. A. Harrington, C. E. Wilkes, Macromolecules, (1977), 10, 536.)

TABLE IV.

Computer Fit of a $^{13}C$ nmr Spectrum
Using Method of Conditional Probabilities

| Type | Calculated | Observed | Difference |
|---|---|---|---|
| S$\alpha\alpha$ | 5.98 | 6.00 | 0.02 |
| S$\alpha\alpha$ | 10.92 | | |
| S$\alpha\delta^+$ | 19.99 | | |
| | 30.91 | 30.00 | -0.91 |
| S$\alpha\beta$ | 4.41 | 4.50 | 0.09 |
| T$\gamma\gamma$ | 0.56 | | |
| T$\gamma\delta^+$ | 2.38 | | |
| T$\delta^+\delta^+$ | 10.27 | | |
| | 13.31 | 14.00 | 0.79 |
| S$\gamma\gamma$ | 3.40 | | |
| T$\beta\beta$ | 1.53 | 14.90 | 1.08 |
| T$\beta\gamma^+$ | 8.89 | | |
| S$\gamma\delta^+$ | 12.45 | 12.70 | 0.25 |
| S$\delta^+\delta^+$ | 26.12 | 27.50 | 1.38 |
| | 52.39 | 55.10 | 2.71 |
| S$\beta\gamma$ | 0.75 | | |
| S$\gamma\delta^+$ | 19.24 | | |
| | 19.99 | 19.00 | -0.99 |
| S$\beta\beta$ | 5.46 | 6.00 | 0.54 |
| P$\beta\beta$ | 1.53 | | |
| P$\beta\gamma^+$ | 8.89 | | |
| P$\delta^+\delta^+$ | 13.21 | | |
| | 23.64 | 21.50 | -2.14 |

TABLE V.

$^{13}$C NMR Analysis of Two EPDM Rubbers

|                    | EPDM A | EPDM B |
|--------------------|--------|--------|
| mol % $C_3H_6$     | 35.7   | 20.1   |
| "$r_1 r_2$"        | 0.78   | 0.60   |
| % inversion        | 8      | 48     |

TABLE VI.

Methylene Sequence Distribution of Two EPDM Rubbers

Percent Methylene in Sequence Lengths 1-12+

|        | 1    | 2    | 3     | 4    | 5     | 6    | 7     | 8    | 9     | 10   | 11   | 12+   |
|--------|------|------|-------|------|-------|------|-------|------|-------|------|------|-------|
| EPDM A | 5.50 | 4.05 | 15.06 | 1.38 | 15.63 | 1.29 | 13.62 | 1.07 | 10.90 | 0.83 | 8.29 | 22.36 |
| EPDM B | 1.55 | 1.06 | 3.23  | 3.50 | 4.23  | 4.12 | 4.64  | 4.31 | 4.68  | 4.23 | 4.29 | 59.96 |

The measurement of methylene sequence length in ethylene propylene rubber has been performed by Randall (13) using a less complicated scheme than the conditional probability method. Ray and coworkers (16) also suggested an analysis for those ethylene propylene rubbers which have no inverted propylene present. The advantage of the statistical model analysis is to test different polymerization models and mechanisms (11) that cannot be tested with a method limited (13) to sequence distributions. Zambelli and coworkers (17,18,19,20,21) have suggested a reasonable model for vanadium-based Ziegler catalyzed ethylene propylene rubbers. Their model has a predominance of secondary insertion for propylene adding to propylene and of primary insertion for propylene adding to ethylene. We expressed the mechanisms of their model in terms of conditional probabilities, which resulted in a model using three parameters instead of five. A comparison of the two models showed little differences in monomer composition; only slight differences in "$r_1r_2$" and inversion, but significant differences in a portion of the spectral fit.

Figure 2 shows the profile of the 27-29 ppm spectral region of three polymers which served as models (11) for ethylene propylene rubber. The better agreement between the observed spectrum and the five-parameter model strongly suggests the three-parameter model is less realistic as an explanation for the polymerization mechanism. Table VII compares the observed profiles of EPDM rubbers made with a Ziegler catalyst system. The ratio of $S_{\gamma\gamma}/S_{\beta\gamma}^+$ (i.e., H/L) agrees better with the five parameter model

## TABLE VII.

### Profiles of 27-29 ppm Spectral Region for Polymerization Models (11)

|       | mol % $C_3H_6$ | CHW-5 | Z-3  | Observed |
|-------|----------------|-------|------|----------|
| J/I   | 25.3           | 2.9   | 3.0  | 3.0      |
|       | 35.5           | 1.9   | 2.0  | 1.9      |
|       | 51.0           | 1.3   | 1.5  | -        |
| H/L   | 25.3           | 0.98  | 2.60 | 0.8      |
|       | 35.5           | 0.51  | 0.81 | 0.5      |
|       | 51.0           | 0.26  | 0.90 | 0.4      |
| L/J   | 25.3           | 0.07  | .02  | .07      |
|       | 35.5           | 0.22  | 0.14 | .27      |
|       | 51.0           | 1.40  | 0.29 | 1.3      |

$$J/I - S\delta^+\delta^+/S\gamma\delta^+$$
$$H/L = S\gamma\gamma/T\beta\gamma^+$$
$$L/J = T\beta\gamma^+/S\delta^+\delta^+$$

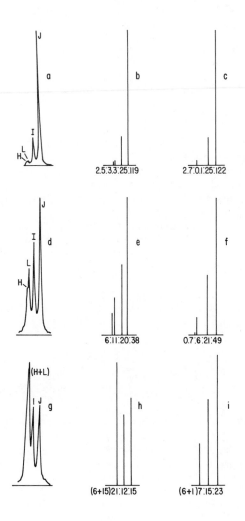

Macromolecules

*Figure 2. Analysis of 27–29 ppm $^{13}C$ nmr spectral region (peaks HIJL) for EPDM samples containing 24, 43 and 53 wt % $C_3H_6$. (a) experimental; (b) calculated using five-parameter model; (c) calculated for parameters simulating Zambelli model (11).*

than the three parameter model over the propylene composition
range cited.

Just as high resolution, solution spectra have yielded much
molecular structure information on ethylene propylene rubbers,
the quality of similar data on other elastomer systems has been
fruitful. Figure 3 shows the $^{13}$C nmr spectra of copolymers of
propylene and butadiene made with both a vanadium and a titanium
catalyst. The two copolymers were hydrogenated and the $^{13}$C nmr
spectra of the resulting polyalkanes showed they had regular
repeating sequences of five methylene carbons bounded by tertiary
carbons bearing methyl groups (22). Based on comparing the
chemical shifts of the polyalkanes shown in Figure 4 to those
seen in ethylene propylene rubbers and to empirical predictions,
the $^{13}$C nmr method was unambiguous in assigning a perfectly
alternating sequence distribution. The extra resonances seen in
Figure 3b for the titanium-made polymer are the result of cis
configuration substituent effect and also the presence of cis-1,4-
polybutadiene homopolymer. The spectra of the original polymers,
shown in Figure 3, show that the copolymer made with a vanadium
catalyst is alternating with essentially all of the butadiene in a
trans configuration. There appears to be less than 2% polybutadiene homopolymer present as an impurity. On the other hand, the
copolymer made with titanium is an alternating copolymer with
about 89% of the butadiene in a trans configuration and 11% in a
cis configuration. This copolymer also had 7-8% cis-1,4-polybuta-
diene present as an impurity.

The quantitative use of high resolution solution spectra for
the determination of configuration has also been ably demonstrated
for other elastomer systems such as polypentenamer (23,24,25),
polybutadiene (26,27), and polyisoprene (1). One would like to
explore the possibilities of analyzing solid elastomer systems in
terms of high resolution spectral discrimination.

High Resolution Spectra of Solids. Figure 5 compares the
spectra of various cis-1,4-polyisoprene samples obtained from the
Fourier transform of Bloch decays using typical proton decoupling
powers typical for small organic molecules. One can see that the
line widths of the cured, carbon black filled rubber are greater
than those of the solution spectra. However, the latter spectrum
is still narrow enough to provide chemical shifts and allow for
material identification. The sample used to obtain the spectrum
in Figure 5c was run using magic angle spinning. The results are
shown in Figure 6. The lines are narrower than that obtained from
the high temperature spectrum shown in 5c. The narrow resonances
obtained with the ambient temperature, magic angle spinning have
the inherent possibility of assigning resonances of minor structures resulting from cross-linking structures or additives.

Caution should be used in assuming an inherent advantage
exists in obtaining a magic angle spectrum of a rubber gum stock.
We found the line widths obtained on uncured polyisoprene with

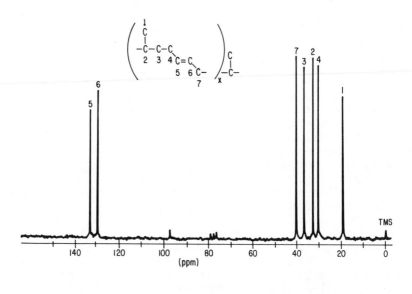

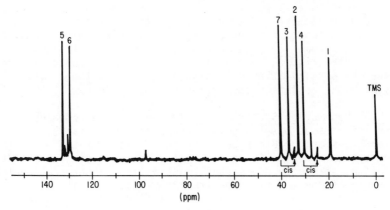

*Figure 3.* C-13{¹H} *NMR spectrum of alternating propylene butadiene copolymer made with a (a) vanadium and (b) titanium catalyst. This was obtained at
ambient temperature from a 20% (w/V) solution in 1:1 CCl₃:CDCl₄ (32).*

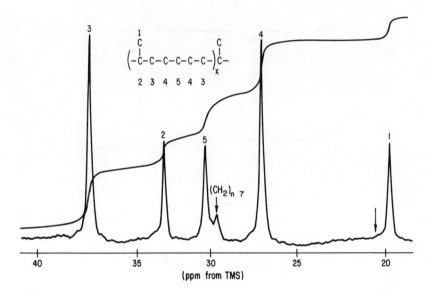

*Figure 4.  Pulsed FT C-13 NMR spectrum of the polyalkane obtained by hydrogenating an alternating propylene butadiene copolymer made with a titanium catalyst, Copolymer B. The spectrum was obtained at ambient temperature from a 20% (w/V) solution in 1:1 CCl₄:CDCl₃ (32).*

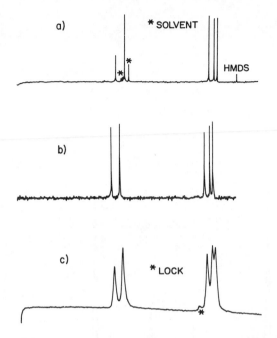

*Figure 5.  C-13{¹H} NMR spectra of cis-1,4-polyisoprene: (a) in perchloroethyl-ene at 90°C; (b) as a solid gum stock at 90°C; (c) as a cured solid with 50 phr carbon black at 100°C.*

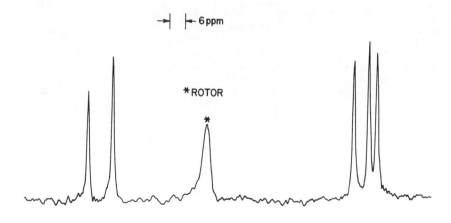

*Figure 6.  Magic angle spinning, high-power proton decoupling, FT C-13 NMR spectrum of cured, carbon-black-loaded polyisoprene at ambient temperature, FT of normal FID without proton enhancement.*

magic angle spinning at room temperature were very narrow.  But
they were no narrower than Figure 5b, a high temperature, normal
pulsed spectrum.
  A cross polarization experiment (29) without magic angle
spinning was attempted and the resulting spectrum is shown in
Figure 7.  This spectrum is consistent with an interpretation (30)
of a superposition of an anisotropic carbon black spectrum and a
spectrum of a natural rubber phase of lower mobility on the filler
surface.  Our attempts to combine the merits of magic angle spinning and cross polarization for this elastomer is a manner similar
to that done with glassy polymers (31) have not been successful.
Difficulties arise because a rotor must be filled with the soft
rubber, rather than fashion the rotor from the polymer (31).
Also, the high mobility present in elastomers creats a weak
dipolar coupling so that the cross polarization is inefficient and
results in weak enhancement compared to standard free induction
decay spectra.  As far as material identification is concerned,
the spectrum resulting from acquiring a standard pulsed free
induction decay at an elevated temperature is adequate.  Further
research will probably show the narrow lines from the magic angle
spectra of natural rubber may allow assignments to lesser components.
  For comparison to polyisoprene, some spectra of other solid
elastomers will be shown to further demonstrate the quality of
solid elastomer spectra.
  We have previously demonstrated (1) that high resolution
spectra of styrene butadiene rubbers (SBR) could be used to distinguish the difference between a solution polymerized SBR and a
blend of SBR with high cis polybutadiene rubber (BR) having the
same overall styrene content.  Figure 8 compares the solution and
solid spectra of a 60/40 blend of SBR and BR.  The identification
of the solid as a blend is evident.  Figure 9a shows that the
resonances are broadened when the elastomer system is cured and
carbon black is present.  Figure 9b shows the same sample run with
magic angle spinning in a Kel F rotor.  Fine structure is beginning to be seen in the aliphatic and aromatic regions and the
smaller olefinic carbons are becoming evident.  As with the natural rubber system, we have not been able to achieve enhanced
sensitivity using cross polarization with these samples.
  Figure 10 shows a spectrum of butyl rubber gum stock obtained
on the solid at 80°C using normal pulsed FT techniques.  Clearly
it could be identified as a component in fabricated materials by
direct $^{13}$C nmr spectral analysis.  Figure 11 shows spectra
obtained from various portions of typical rubber products.  These
samples were cut from the rubber product, placed in an nmr tube
without solvent, and spectra obtained at an elevated temperature.
The data show how polyisoprene, a polyisoprene/polybutadiene blend
and a polyisobutylene/polyisoprene/polybutadiene rubber blend are
quickly identified in the materials.  Figure 11a shows processing
oil was present, and which was confirmed by solvent extraction.

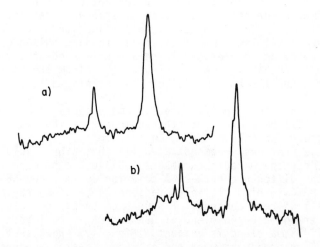

*Figure 7. Cross polarization spectra of cured, carbon-black-filled polyisoprene using (a) spin lock of 20 msec, contact time 10 msec and (b) spin lock of 20 msec, contact time 5 msec.*

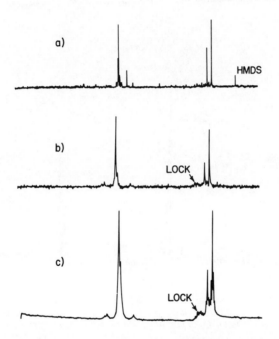

*Figure 8.* *C-13 NMR spectra of a 60/40 emulsion SBR/polybutadiene rubber blend: (a) in perchloroethylene at 100°C; (b) solid, uncured unfilled rubber at 90°C; (c) solid, cured unfilled rubber at 100°C.*

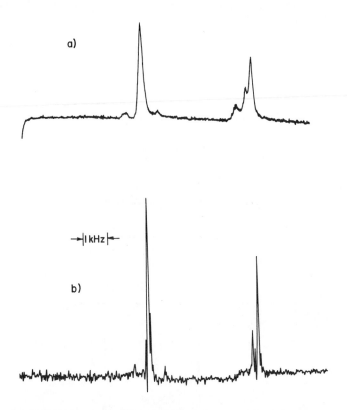

**Figure 9.** *C-13 NMR spectra of a 60/40 emulsion SBR/polybutadiene rubber blend: (a) solid, cured, carbon black filled at 100°C; (b) same as Sample a but using magic angle spinning at ambient temperature.*

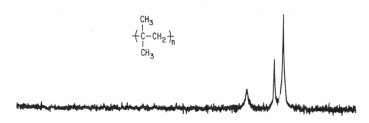

**Figure 10.** *C-13 NMR spectrum of butyl rubber as a solid at 70°C*

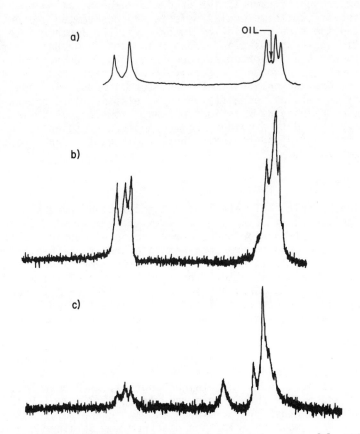

*Figure 11.* *C-13 NMR spectra rubber fabricated materials run on solids at 90°C using normal pulsed FT technique: (a) tire tread; (b) tire compound; (c) liner material.*

## Summary

By examining sections cut from various areas of finished fabricated rubber products, one can easily identify the elastomer component by using standard Fourier transform pulsed nmr techniques combined with high temperature analysis. Such an analytical approach is useful for comparing product uniformity or for the comparison of competitive products. For detailed understanding of subtle molecular defects and associated properties or for studying rubber at filler interfaces, the further development of magic angle spinning and cross polarization techniques for elastomers is anticipated as necessary. Until these combined techniques are used more extensively with practical elastomer systems, one can only estimate their usefulness. The spectral resolution obtained directly on solid elastomers is not as great as the corresponding solution spectra. However, a combination of both types of data provides the macromolecular scientist with a new dimension in understanding both the structure and properties of polymers.

## Experimental

The experimental details for the solution spectra have been previously given (1).

The normal Fourier transformed proton decoupled spectra of the solid elastomers were obtained on a Bruker HX-90-E/SXP spectrometer with a 15-inch magnet. The solid samples were cut into small pieces and placed into the inner portion of a 7 mm ID Wilmad coaxial tube. The upper portion of the inner coaxial tube was 9 mm ID and fits precisely inside a standard 10 mm Wilmad tube which contained the high temperature lock compound. Depending on the desired temperature, either $D_2O$, $DMSO-d_6$ or 1,4-dibromotetradeuterobenzene was used as the lock. Ninety degree pulse widths of less than 15 μs were used with a five second repetition time, 16 K data points and 6 KHz sweep width. The spectra combined with magic angle spinning were obtained courtesy of Bruker Instruments using a Bruker CXP pulsed spectrometer at 45.3 MHz, 1.5 KHz spinning rates, 6 gauss $H_1$ for proton and a $^{13}C$ $\pi/2$ pulse of about 8.5 μs. The cross polarization spectrum was obtained courtesy of Dr. J. Schaefer, Monsanto Research, using a spin lock of 20 ms, a single contact of 5 ms and 10 ms and a $H_1$ for proton of about eight gauss.

The polymer samples were standard commerical rubbers.

## Abstract

High resolution $^{13}C$ nmr spectroscopy of elastomers is a powerful tool for identification and characterization of elastomer systems. The $^{13}C$ nmr spectra of elastomers in solution are rich in molecular structure details. The information provided

can be in the form of monomer sequence distribution, chain configuration, steric purity or identification of polymer mixtures.
Comparison of $^{13}$C spectra of solid elastomers with those from
solutions shows that structural information can also be obtained
on the solids. Spectra of elastomer solids under conditions of
normal free induction decays, cross polarization and magic angle
spinning will be described. A few applications to some typical
elastomer composites will be discussed.

## Literature Cited

1. Carman, C. J. and Baranwal, K. C., Rubber Chem. Technol.,
   (1975), 48, 705.
2. Duch, M. W. and Grant, D. M. Macromolecules, (1970), 3, 165.
3. Schaefer, J., Chin, S. H., and Weisman, S. I.,
   Macromolecules, (1972), 5, 798.
4. Carman, C. J., and Wilkes, C. E., Rubber Chem. Technol.,
   (1971), 44, 781.
5. Wilkes, C. E., Carman, C. J., and Harrington, R. A.,
   J. Polym. Sci., (1973), 43, 237.
6. Crain, W. O., Jr., Zambelli, A., and Roberts, J. D.,
   Macromolecules, (1971), 4, 330.
7. Zambelli, A., Gatti, G., Sacchi, C., Crain, W. O., Jr., and
   Roberts, J. D., Macromolecules, (1971), 4, 475.
8. Tanaka, Y., and Hatada, K., J. Polym. Sci., Polym. Chem. Ed.,
   (1973), 2057.
9. Grant, D. M., and Paul, E. G., J. Amer. Chem. Soc., (1964),
   86, 2984.
10. Carman, C. J., Tarpley, A. R., Jr., and Goldstein, J. H.,
    Macromolecules, (1973), 6, 719.
11. Carman, C. J., Harrington, R. A., and Wilkes, C. E.,
    Macromolecules, (1977), 10, 536.
12. Van Schooten, J., and Mostert, S., Polymer, (1963), 4, 135.
13. Randall, J. C., Macromolecules, (1978), 11, 33.
14. Lindsay, G. A., Singleton, C. J., Carman, C. J., and Smith,
    R. W., Polymer Preprints, (1978), 19(1), 206.
15. Carman, C. J., Batiuk, M., and Herman, R. M., U. S. Patent
    4,046,840 (1977), assigned to BFGoodrich.
16. Ray, G. J., Johnson, P. E., and Knox, J. R., Macromolecules,
    (1977), 10, 773.
17. Bovey, F. A., Sacchi, M. C., and Zambelli, A.,
    Macromolecules, (1974), 7, 752.
18. Zambelli, A., Wolfsgruber, C., Zannoni, G., and Bovey, F. A.,
    Macromolecules, (1974), 7, 750.
19. Locatelli, P., Provasoli, A., and Zambelli, A., Makromol.
    Chem., (1975), 176, 2711.
20. Wolfsgruber, C., Zannoni, G., Rigamonti, E., and Zambelli,
    A., Makromol. Chem., (1975), 176, 2765.
21. Zambelli, A., Tosi, C., and Sacchi, C., Macromolecules,
    (1972), 5, 649.

22.  Carman, C. J., Macromolecules, (1974), 7, 793.
23.  Carman, C. J., and Wilkes, C. E., Macromolecules, (1974), 7, 40.
24.  Chen, H. Y., J. Polym. Sci., Lett. Ed., (1974), 12, 85.
25.  Ivin, K. J., Laverty, D. T. and Rooney, J. J., Makromol. Chem., (1978), 179, 253-258.
26.  Clague, A. D. H., van Broekhoven, J. A. M., and Blaauw, L. P., Macromolecules, (1974), 7, 348.
27.  Elgert, K. F., Quack, G., and Stutzel, B., Makromol. Chem., (1975), 176, 759.
28.  Schaefer, J., Stejskal, E. O., and Buchdahl, R., Macromolecules, (1977), 10, 384.
29.  Pines, A., Gibby, M., and Waugh, J. S., J. Chem. Phys., (1973), 59, 569.
30.  Schaefer, J., High Resolution $^{13}$C nmr Studies of Solid Polymers, "Molecular Basis of Transition & Relaxation", Gordon & Breach Publishers, U.K., 1978, p. 103.
31.  Schaefer, J., Stejskal, E. O., and Buchdahl, R., J. Macromol. Soc. Phys., (1977), B13(4), 665.

32.  Carman, C. J., Macromolecules, (1974), 7, 789.

Discussion

J. Prud'homme, U. of Montreal, Que.: Did you consider the possibility of studying swollen rubbers (vulcanized materials) to obtain better resolution?

C. J. Carman: There is no advantage. We did not obtain lines any narrower than when the samples were run at higher temperatures. Solvent was excluded because we were also trying to determine effects of relaxation times. The effects are hard to interpret. Solvent present would cause even greater difficulties in the interpretation of the relaxation data. As far as the high resolution quality of the spectra, we see no advantage in having solvent present in these elastomer systems. There is an advantage in swelling a plastic system because mobility of the chains will be affected and higher resolution achieved.

J. Guillet, U. of Toronto, Ont.: It seems to me that when one talks about solids one needs to define what is meant by a solid. In this NMR measurement one is looking at something similar, in terms of its molecular properties, to the kinds of motions necessary for the diffusion of small molecules through polymers. The conventional measurement of polymer viscosity looks at the probability of motion of the center of mass of a very large molecule, which nevertheless is still very small indeed. This leads to very large viscosities for the system. A small molecule moving through a polymer will be assisted by the small scale motions (such as, rotations or vibrations of chains). These latter are the sorts of mobility examined by NMR. Probably the fact that you are dealing with a rubber may simply be that some of the restrictions have been eliminated that would be present if you had a semi-crystalline polymer. This brings me to my

question.  Polyethylene for example or an ethylene-propylene
rubber will have a glass transition of about -40°C.  Are the same
results obtained when a solid sample of polyethylene is subjected
to these high temperatures as when a strictly amorphous material
such as polyisoprene or some of the other rubbers you talked
about are run at high temperatures?

    C. J. Carman:  I agree whole-heartedly with the first part
of your statement.  We have to define what we mean by a solid and
that rubbers or elastomers on an NMR time scale really have a lot
of mobility and should not be thought of as solids.  This is one
reason I presented these data:  to stimulate discussion.  There
is a school of thought that all polymers should be classified as
solids and therefore solids techniques should be applicable.
Because differences do exist between "solids" I agree with the
first part of your statement.  With regard to your second point,
differences are evident.  Spectra of solid EPDM exhibit broader
peaks, as one would expect, than do polymers such as polyisoprene.
I am not willing to say a whole lot about the advantage of the
cross polarization and the magic angle spinning because I found
the peaks do not narrow as you just predicted they should and I
would predict they would.  I think there is still some work to be
done.  Even though there is some crystallinity present, I am not
certain that we can define the crystallinity as measured by x-ray
diffraction in terms of the motions and the constraints that bear
on the cross polarization experiment.  I can't define it.  We
have not seen good cross polarization effects on these samples
and the magic angle spinning didn't narrow the peaks that much
more than does a higher temperature.  High temperatures do not
give as narrow a peak as found in solution spectra of, for
example, polyisoprene.

    J. Guillet:  Can you relate the broadness to the crystallin-
ity or are you at a temperature well above the melting point?

    C. J. Carman:  We are at a temperature well above the melt.

    J. Guillet:  The thing that is not realized is that diffus-
ion of benzene through rubber is about the same as benzene
through benzene.  The kinds of motion found in a liquid are in
fact very similar to those found in a rubber.  It's just that the
translational motion of the large molecules is restricted.  It is
not too surprising that rubbers would show up in some cases as
liquids.

    C. J. Carman:  Yes, I agree.

    J. Guillet:  The swelling doesn't help you very much because
it hasn't affected the mobility of the molecules.  It has helped
the macroscopic appearance of viscosity but not the microscopic.

    C. J. Carman:  Agreed.  But as far as those ethylene-pro-
pylene polymers which have crystallinity and a Tg of around minus
forty or fifty degrees, I expect they might be in a different
class.  Here there might be some advantage to swelling.

    A. Garroway, Naval Research Lab, Washington, DC:  A few
quick comments.  First of all, it's worth reminding ourselves that

the NMR definition of a liquid looks at a specific dipolar or NMR
type interaction and asks whether it is averaged out.  Depending
on what interactions are involved, what molecules are being
examined, what experiment is being carried out and what nuclear
species is being measured, one may have simultaneously a liquid
and a solid preventing characterization of materials on a rheo-
logical basis.  One can have a partial system.  A methyl rotation
would be a classic example where one has very fast anisotropic
motion.  It is necessary therefore to be very careful about applying entropomorphic notions of what a liquid or solid is.  Secondly,
a comment about cross polarization.  It is in fact possible to do
cross polarization on pure liquid type systems and by that I mean
something in which all dipolar interactions are averaged out.
There one uses the J interaction to couple and the advantage that
one obtains is that there is some enhancement of signal to noise.
There is also a speed-up in time.  The experiments formally look
quite different because no equilibration of temperature exists
but rather an oscillation which goes on continuously.  So I agree
with everything said but ask that cross polarization not be
indited just yet.

C. J. Carman:  Actually, I didn't intend to.  I'm glad you
made the comment you did.

W. G. Miller, U. of Minnesota, MN:  My question has reference to your cured systems which were filled with carbon black.
I would have thought that a reasonable fraction of the polymer
was very much immobilized with the carbon black.  If so why do
you not see it when cross polarization experiments with spinning
are carried out?

C. J. Carman:  We did try to look at the bound rubber.  It
turns out that there still is too much motion in such rubber to
get cross polarization.  You can see it by bound rubber measurements.  On the time scale of this experiment it does not contribute to the spectrum.

W. G. Miller:  How do you know you are not seeing bound
rubber?

C. J. Carman:  Let me answer this way.  We know there is
bound rubber present using standard techniques.  I have yet to see
the bound rubber with the $^{13}$C NMR technique, is what I am saying.
Area measurements of high resolution spectra do not give numbers
for bound rubber that coincide with standard measurements.  For
this reason I am not convinced I have ever seen it with NMR.

J. McAndless, Defense Research, Ottawa:  The C-13 NMR technique seems quite good for identifying the major component in the
rubber formulations, namely the rubber itself.  Is the technique
sufficiently sensitive to pick out the antioxidants and the processing oils without having to go through the normal separation
techniques?

C. J. Carman:  In some cases, yes.

J. McAndless:  Does it require use of a differential technique computer based program to subtract out the C-13 NMR rubber

signals?

C. J. Carman: I would like to ask the audience if anybody has been successful in doing NMR subtraction spectroscopy? We haven't been that successful at these very low levels because of artifacts. The appropriate signals can be seen directly without resorting to subtraction.

D. Axelson, Florida State Univ. FL: In the light of what I am going to talk about tomorrow I would like to ask whether you have any numbers for the bulk polyethylene or the bulk ethylene copolymer line widths?

C. J. Carman: I have them but not with me.

D. Alexson: Basically are we talking about 50 Hz, 100 Hz, 300 Hz...?

C. J. Carman: They are on the order of 25-50 Hz. I have to qualify that because most of our measurements are at high temperatures. So when you talk about the bulk, one should specify at what temperature, at what frequence?

D. Axelson: These line widths are for 100°C at 22 MHz?

C. J. Carman: The line widths for EPDM are on the order of 25-60 Hz at 100°C at 22 MHz. They will vary a little.

D. Axelson: Are these high molecular weight polymers?

C. J. Carman: Yes.

J. Guillet: It intrigues me that the EPDM exhibit broadening. I'm wondering whether it is crystalline order that is being observed above the melting point. Is there truly random polymer above the melting point? Would this be a possible explanation for the broadening?

C. J. Carman: I guess it could be, but again I'm not certain because there will be a distribution of chemical shifts which perhaps contributes to the line widths as well. In addition to the latter, other factors can contribute to line broadening. As was indicated by other speakers before me, conformational effects do play a role in polymer spectra, I'm convinced of it. I'm not certain of how to extract it but I'm convinced it is there. In the center of the EPDM spectrum is the bulk of the structure which arises from the long runs of methylene carbons. There are several peaks which are very temperature dependent and very solution dependent. I am not sure how much of the line width is really due to molecular motion as opposed to distribution of chemical shifts. The shifts may differ from those found for the polymer when it is in solution. This is why I am a little hesitant to assign the broadening to crystallinity effects.

RECEIVED March 13, 1979.

# Characterization of Carbohydrate Polymers by Carbon-13 NMR Spectroscopy

ARTHUR S. PERLIN and GORDON K. HAMER

Department of Chemistry, McGill University, Montreal, Quebec, H3A 2K6, Canada

Many of the advantages offered by [13]C-NMR spectroscopy for studies on macromolecules in general, are represented in applications to carbohydrate polymers. Although for a number of years [1]H-NMR studies have furnished valuable information in the field -- early applications to dextrans (1) and glycosaminoglycans (2) may be cited as examples -- and continue to do so, a serious limitation is imposed by the fact that [1]H-signals of the polymers are often excessively broad, even in spectra recorded at high field. Usually, this problem is less serious with [13]C-NMR, and the wide range of chemical shifts (> 200 p.p.m.) for [13]C nuclei is another characteristic that favors the resolution of most of the individual signals in a spectrum.

A comparison of spectra of hyaluronic acid (3,4) furnishes an illustration of the advantage frequently offered by [13]C-over [1]H-NMR spectroscopy. As seen in Fig. 1, the proton spectrum at 220 MHz conveys little information about the fact that the polymer is composed of a disaccharide repeating sequence (1). By contrast, fourteen signals are evident in the [13]C spectrum, thus accounting for each nucleus of sequence 1. It is worth noting that both spectra in Fig. 1 were recorded at an elevated temperature so as to reduce line broadening, although the improvement in resolution obtained was greater for the [13]C spectrum. Because of this feature, as is well known (5,6), [13]C-NMR spectroscopy can be advantageously employed in studying phase transitions of polymers such as hyaluronic acid (4), and other polysaccharides (7,8).

Characteristic Chemical Shifts.

Fig. 1 also may be used to note some general characteristics of [13]C spectra of carbohydrate polymers (9-11). Chemical shifts of anomeric carbons (C-1), in the region of 100-110 p.p.m., are typically well separated from other signals. As compared with C-1 of the related monosaccharides (12-15), the anomeric carbon is strongly deshielded (commonly by 7-10 p.p.m.) through glycoside formation (9), i.e., by the change from O-H to O-C bond.

0-8412-0505-1/79/47-103-123$05.00/0

Similarly, the C-1 resonance of an axial anomer is shielded relative to that of its equatorial isomer. Also very distinctive are
signals due to the carbon of a primary alcohol group (C-6, in the
region of 60-65 p.p.m.) and to the carboxyl group of an uronic
acid moiety. Typically, as seen in Fig. 1, the carboxyl C=O resonance is in the region of 175 p.p.m. although, as noted below (see
Fig. 3), it is strongly pH dependent.

Usually, the carbon involved in glycosidic bonding to the anomeric position of the adjacent residue is sufficiently deshielded by this bond as to produce a signal well separated from those
of the other classes of carbons. Hence, in Fig. 1, the two signals near 80 p.p.m. are assigned to C-4 of the two residues in 1.
The remaining, secondary, carbons have chemical shifts commonly
centred around 75 p.p.m. Their signals constitute the most difficult ones to assign, although in certain applications this need
not be a serious deficiency. A generally useful approach (9-11,
16,17) to an analysis of these signals involves correlations with
spectral data for model compounds -- monosaccharides and oligosaccharides, and derivatives, as well as other polysaccharides.
Some examples of chemical shift correlations of this type are
given later. Other techniques include selective [1]H-decoupling,
and chemical or enzymatic modification. It is fair to say, however, that the discovery of new approaches to this problem could
very materially facilitate fuller analyses of the spectra.

The particular array of chemical shifts found for the [13]C
nuclei of a given polymer depends, of course, on such factors as
bond orientation, substituent effects, the nature of nearby functional groups, solvation influences, etc. As a specific example,
derivatives of the carbohydrate hydroxyl moieties may give rise
to chemical shifts widely different from those of the unmodified
compound, a fact that has been utilized, e.g., in studies (18) on
commercially-important ethers of cellulose. Hence, as illustrated
in Fig. 2, the introduction of an O-methyl function causes (14,15)
a large downfield displacement for the substituted carbon. This
change allows for a convenient, direct, analysis of the distribution of ether groups in the polymer. Analogously, carboxymethyl,
hydroxyethyl and other derivatives may be characterized as well
(18).

## Differentiation of Polymers. Mucopolysaccharides

A program in our laboratory on the chemistry of mucopolysac-
charides (glycosaminoglycans) has been greatly  enhanced by the
availability of [13]C-NMR. For instance, its application (10,19,
20) to studies on the blood anticoagulant, heparin, has served
to re-inforce evidence from [1]H spectra (2) showing that there are
two main classes of heparin. One type is exemplified by material
extracted from beef lung (B type), which may be depicted almost
wholly by a repeat biose structure (2). A second type (A type),
from hog mucosa, is shown to contain 15-30% of constituents other

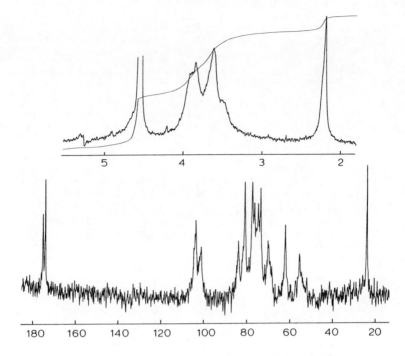

*Figure 1.* ¹H-NMR spectrum at 220 MHz (upper) and C-13 NMR spectrum at 22.63 MHz (lower) of hyaluronic acid (sodium salt) in $D_2O$ solution. Descriptions of analogous spectra are given in Ref. 3 (¹H) and Ref. 4 (C-13).

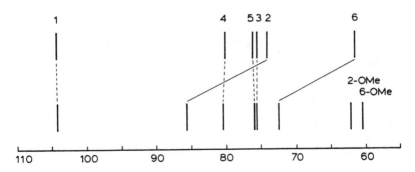

*Figure 2.* Stick diagram representing the C-13 spectra (18) of methyl β-cellobioside, as a model for cellulose (upper), and a partially substituted O-methylcellulose (2- and 6-O-methyl) (lower). The light lines emphasize the changes in chemical shift associated with the introduction of ether substituents.

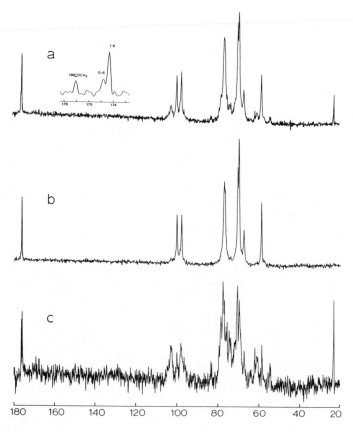

*Figure 3.   C-13-NMR spectra at 22.63 MHz of heparin A (a) and heparin B (b) as sodium salts in $D_2O$ solution. The inset of spectrum (a) shows the carbonyl region of the heparin A spectrum at pH 2. The difference spectrum (c) represents a substraction of spectrum b from spectrum a: I = iduronic, G = glucuronic.*

than those in 2. The most prominent evidence for this is the occurrence of an upfield signal (Fig. 3) attributable to an acetamido methyl carbon, a C-1 signal slightly to low field of those
of the two main constituents, and at least two relatively weak
C=O signals overlapping the main carboxyl peak. However, most of
the signals due to these minor constituents are obscured by the
prominent ones of the main structural components (2). To extract
more information, a difference spectrum has been obtained by sub-
tracting spectrum 3b from 3a, thus eliminating signals of sequence 2 common to both spectra: this gives the pattern shown in
Fig. 3c. The latter bears a striking resemblance to the spectrum
(21) of a major fraction of a related polymer, heparan sulfate,
which consists mainly of acetamido derivative 2a, and β-D-gluc-
uronic acid (2b). Hence, the difference spectrum strongly indicates that these same sugar moieties are the minor constituents of
heparin A, and are bonded to adjacent residues in the polymer in
an analogous manner as in heparan sulfate. These findings accord
well with evidence from several other sources.

Chemical analyses have shown that the compositions of mucosal
heparins vary widely. From results such as those of Fig. 3, it is
clear that [13]C-NMR spectroscopy offers an attractive, rapid
means for the characterization of heparins from different sources,
a long-standing need in the pharmaceutical field.

An illustration of the power of [13]C-NMR for distinguishing
between closely-related carbohydrate polymers, as well as for detecting polydispersity, is provided (22) by studies on chondroitins A, B and C. Although [1]H-NMR finds useful application in this
area, the [13]C spectrum of each of these polymers is more easily
recognizable, and more informative. Very striking is the downfield position of the C-6 signal of the hexosamine residue in
chondroitin C (3c), due to the presence of the 6-sulfate group
(Fig. 4c). In the spectrum of the 4-sulfate (chondroitin A) (3a),
by contrast, the major C-6 signal is found in the region typical
of an unsubstituted primary alcohol group (Fig. 4a). There is
clear evidence from the spectra, moreover, that each polymer contains structural elements of the other, to the extent of 20-30%
(22).

### Chemical Shift Correlations.

Both chondroitins A and C have residues of 4-linked β-D-
glucuronic acid in common (3a and 3c), and there is a close cor-
respondence (22) between the chemical shifts for these residues
in the two polymers (Fig. 4a and 4c). Hence structural differences
in the hexosamine residues of 3a and 3c, give rise to only minor
shielding changes in these acid moieties. A similar analogy is to
be found on comparing the hexosamine 4-sulfate residue common to
chondroitins A and B (3a and 3b). Here again, the corresponding
carbons in the two polymers are characterized by essentially the
same chemical shifts (Fig. 4a and 4b).

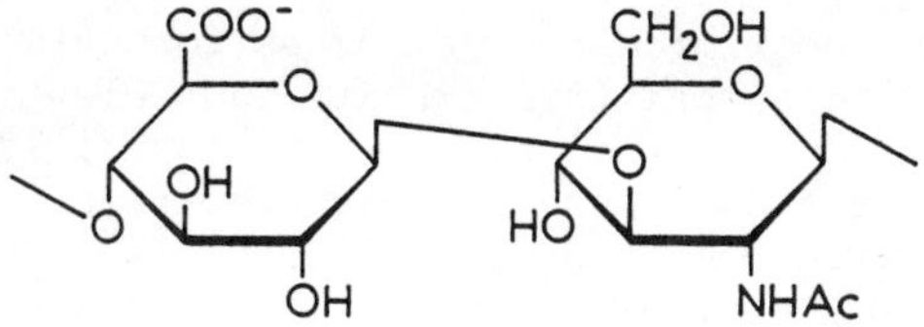


1

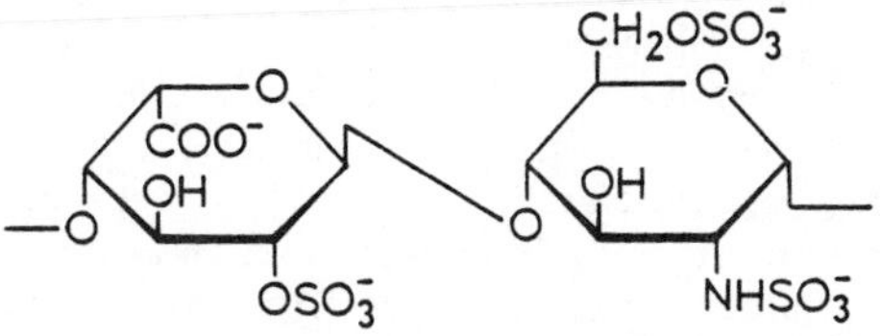


2

2a

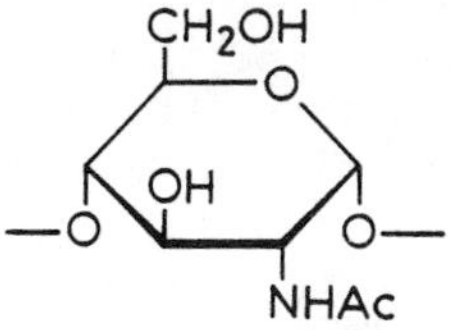


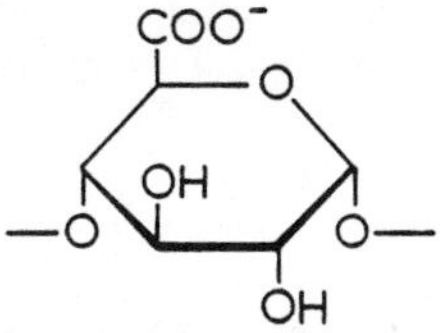


2b

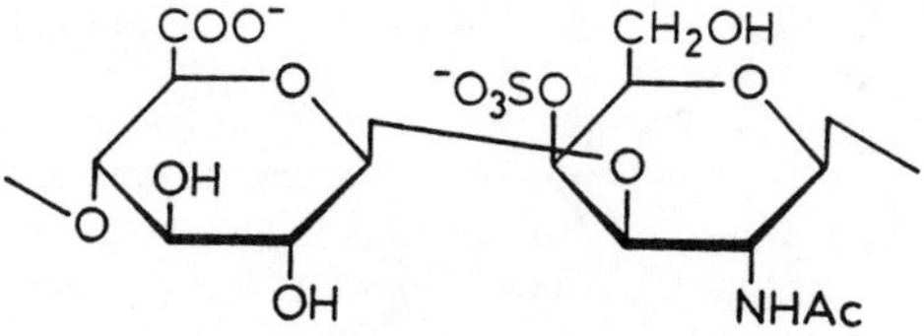


3a

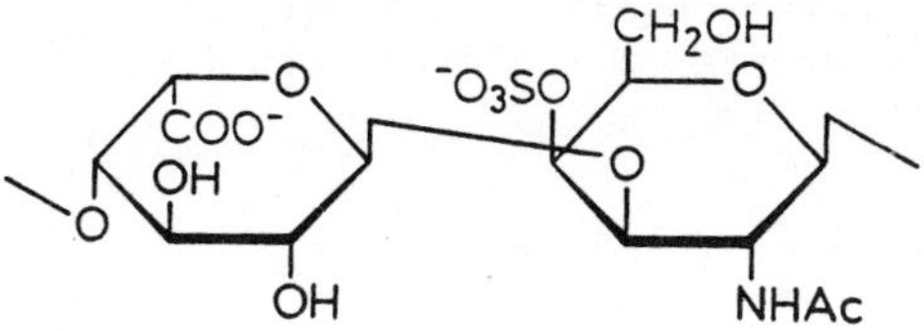


3b

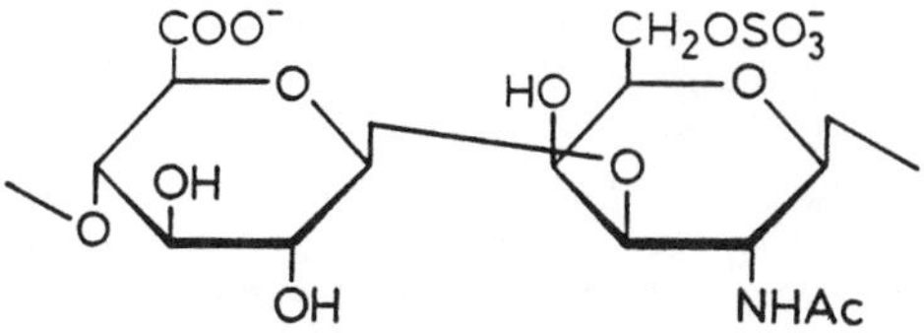


3c

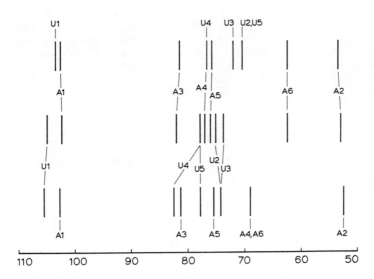

*Figure 4. Stick diagrams representing C-13 spectra at 22.63 MHz (22) of chondroitins A, B, and C (4a, 4b and 4c, respectively). The vertical, light lines relate resonances for analogous C-13 nuclei in the three different polymers. Not included are signals caused by the acetamido $CH_3$ and $C{=}O$ carbons, and the carboxyl carbons (U-6). Minor signals that demonstrate the presence of chondroitin A in C, and of C in A, are found in Figure 1 of Ref. 22. A = acetamideoxyhexose, U = uronic acid.*

It is noteworthy that most of the chemical shift values for
all three polymers may be closely approximated (22) by calculations based on data for monomeric reference compounds. These
findings illustrate, therefore, the general validity of studies
on low molecular weight model compounds for analysis of [13]C spectra of carbohydrate polymers. Many examples of equally satisfactory comparisons of this kind are to be found in studies on other
polysaccharides (11,23). These polymers include glucans (16),
mannans (24,25), limit dextrins (26), lichenin (27), agarose (28)
and various polysaccharides of fungal and microbial orgins (e.g.,
7,8,29-31). Observed departures from expectation have been attributed to specific conformational influences (8).

The studies just cited, and others, serve to illustrate various aspects of the applications of [13]C-NMR described above for
determination of structure, analysis of constituents, detection
of heterogeneity, etc. In this limited review of the subject, unfortunately, it is feasible to deal with only one or two of
those many contributions in order to highlight a particular us-
age.

Enzymic Applications. Agarose.

Studies on agarose and related seaweed polysaccharides, provide an example of how [13]C spectroscopic measurements may be
combined with the use of selective polymer-degrading enzymes for
structural elucidation. In turn, the spectral measurement can become a highly effective means for monitoring characteristics of
the enzymic reactions. The degradation of agarose (4a) may occur
either by cleavage of the β-(1→4)-linkage, giving a series of neo-
agaro-oligosaccharides (4b) having D-galactose at the reducing
end, or by cleavage of the α-(1→3)-linkage, giving agaro-oligosaccharides (4c) with 3,6-anhydro-L-galactose (in the open-chain,
aldehydo form) at the reducing end. The two types of hydrolysis
are clearly differentiated by the [13]C-NMR spectra of crude enzyme
digests (Fig. 5) (28). Cleavage of the β-(1→4)-linkage is indicated by the appearance of new anomeric signals at 93.8 and 97.8
p.p.m. (G'-1α, G'-1β) with intensities in the ratio ∿1:2 (Fig.
5b). In contrast α-(1→3)-cleavage is indicated by the appearance
of a signal at 91.4 p.p.m. (A'-1), a chemical shift characteristic of hydrated aldehydes (Fig. 5c). The spectra show that the
mode of action of each enzyme is specific; there is no evidence
of α-cleavage by β-agarase, or of β-cleavage by α-agarase. Integration of the anomeric carbon peak areas provides an estimate of
the average chain length of the oligomer mixture. For purified
oligosaccharides this may be expressed as n, the number of biose
units per oligosaccharide molecule. By examining the systematic
variations in the [13]C spectra of homologous series of oligosaccharides (4b, 4c; n = 0,1,2) together with the results of model
compound studies, the peak assignments shown in Fig. 5a can be
made. These assignments provide a starting point for determining

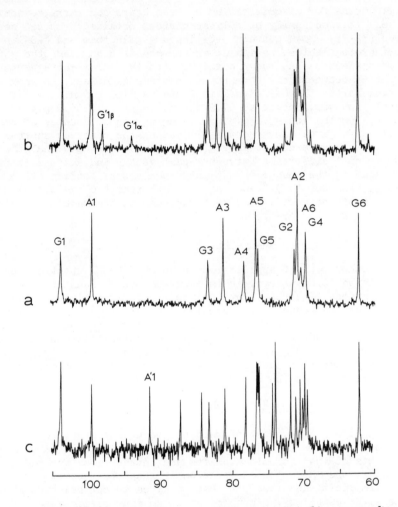

*Figure 5.   22.63-MHz C-13-NMR spectra of: (a) agarose (4a); (b) neoagaro-oligo-saccharides (4b, n = 3.9) produced by enzymic hydrolysis of 4a with β-agarase; and (c) agaro-oligosaccharides (4c, n = 2.5) produced by enzymic hydrolysis of 4a with α-agarase. Spectra recorded in D₂O solution at 35°C (5b,5c) or 95°C (5a). G = D-galactose, A = 3,6-anhydro-L-galactose.*

the presence and location of substituents (e.g., $-OCH_3$, $-OSO_3^-$) in agars isolated from different seaweed species (32,33).

It has also been found that $^{13}C$-NMR spectroscopy provides a facile means for discriminating between agars and carrageenans, a closely related family of polysaccharides obtained from red sea-weeds (32-35). Both contain 3-$\underline{O}$-linked $\underline{D}$-galactose and 4-$\underline{O}$-linked 3,6-anhydrogalactose residues; in carrageenans however, the 3,6-anhydrogalactose has the $\underline{D}$ configuration ($\kappa$- and $\iota$-carrageenan, $\underline{5a}$ and $\underline{5b}$ respectively, are representative examples). In each polysaccharide the G-1 signal is found at $\sim$103 p.p.m. The A-1 signal, however, occurs at 99.2 p.p.m. (agarose), 96.2 p.p.m. ($\kappa$-carrageenan) and 93.1 p.p.m. ($\iota$-carrageenan). Studies of par-tially desulfated carrageenans (36) using the spectrum subtrac-tion technique described above (see Fig. 2), indicate that the 3 p.p.m. shift difference between $\kappa$-carrageenan and agarose is prim-arily due to the change in 3,6-anhydrogalactose configuration, from $\underline{L}$ (agarose) to $\underline{D}$ ($\kappa$-carrageenan) whereas the additional 3 p.p.m. shift between $\iota$- and $\kappa$-carrageenan is the result of sulfa-tion at A-2.

## Monitoring Chemical Change.

Another type of application in which $^{13}C$-NMR spectroscopy may be employed to advantage, is in carrying out various chemical modifications of carbohydrate polymers. It is highly effective for monitoring the reduction of uronide residues ($\underline{20}$), deaminative degradation of aminosugar-containing polymers ($\underline{10}$), acid hydrol-ysis ($\underline{18},\underline{37}$), derivatization ($\underline{18},\underline{23}$), de-esterification; also com-plex formation ($\underline{38}$), interactions with ionic species ($\underline{20},\underline{39},\underline{40}$), etc. In a number of such instances, the spectroscopic measurement may be regarded as the analytical method of choice.

## Uses of $^1H$-coupled Spectra.

Experimentally, most of what has been described above invol-ves the detection of individual $^{13}C$ resonance signals, and mea-surements of their chemical shifts and intensities. For these purposes, the $^1H$-decoupled spectrum is ideal because of its rela-tive simplicity and the fact that it gives an optimal signal re-sponse. Nevertheless, a coupled spectrum can provide valuable in-formation as well ($\underline{10}$). Its simplest use lies in the character-ization of classes of carbon -- methyl, methylene, etc. -- from the multiplicity of the observed signal, in which case off-reson-ance decoupling is often employed. In addition, however, since the magnitude of $^{13}C$-$^1H$ coupling across one, two or three bonds is geometry dependent ($\underline{12},\underline{41}$) stereochemical information may be obtained. Given the quality of $^1H$-coupled spectra now generally available, only measurements of $^1J_{C-H}$ are of practical value with polymers of moderate-to-high molecular weight. Signals of anomer-ic carbons are of particular interest because the magnitude of

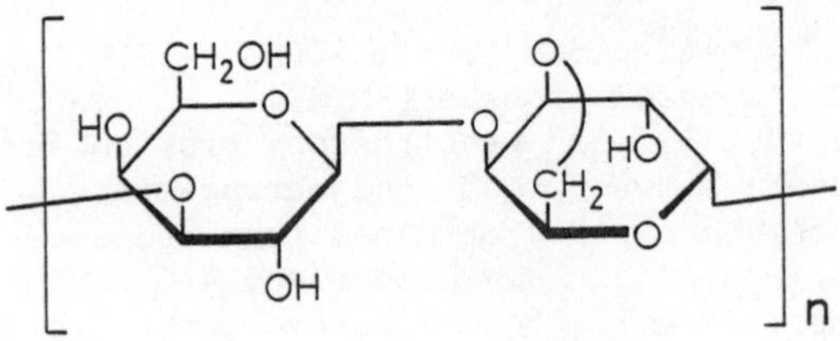

4a

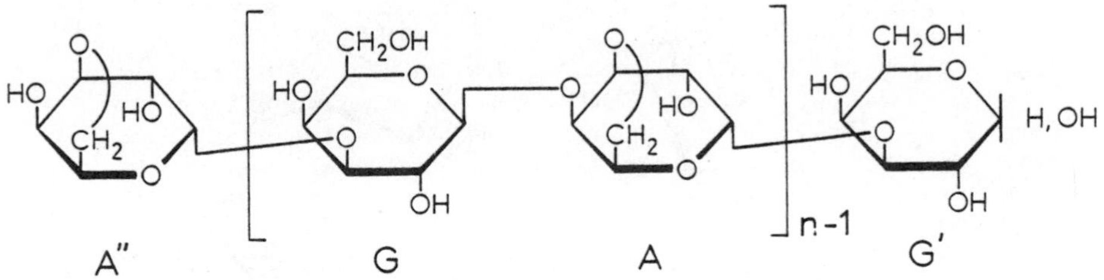

A″        G        A        G′

4b

G″        A        G        A′

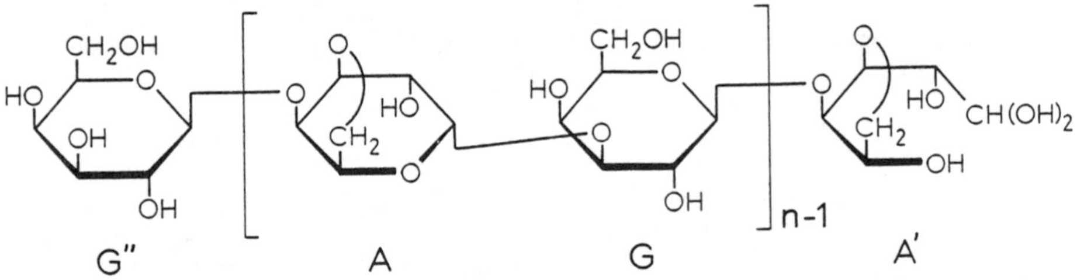

4c

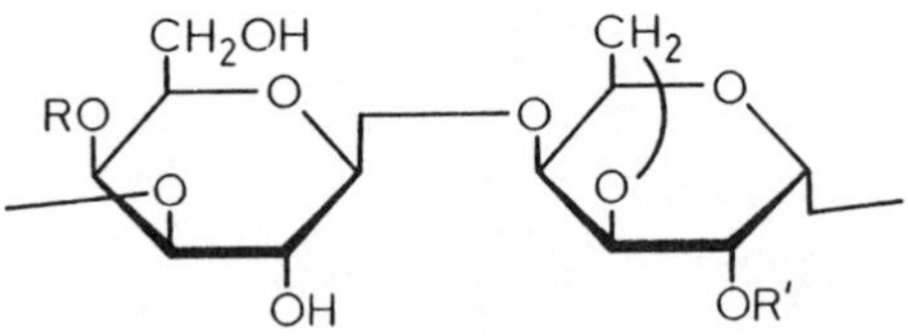

5a  R = SO₃⁻, R′ = H
5b  R = R′ = SO₃⁻

6a

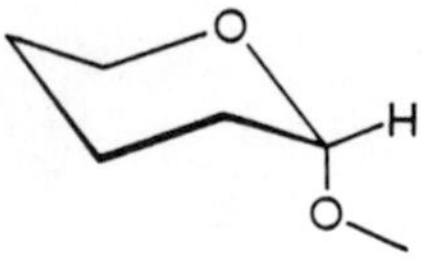


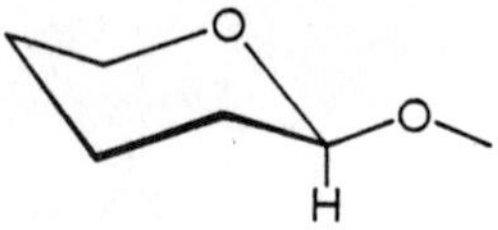


6b

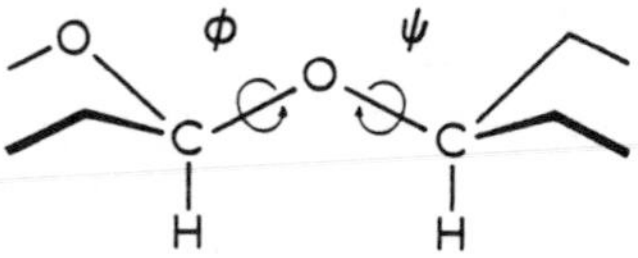


7

8

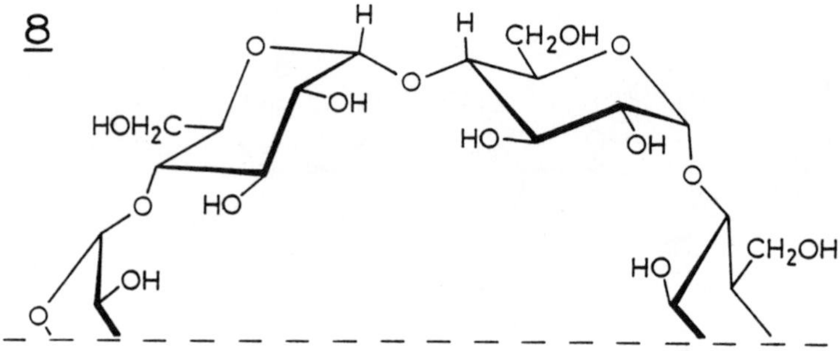


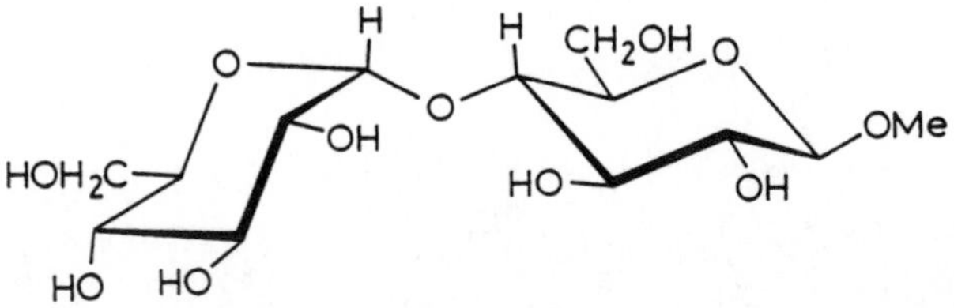


9

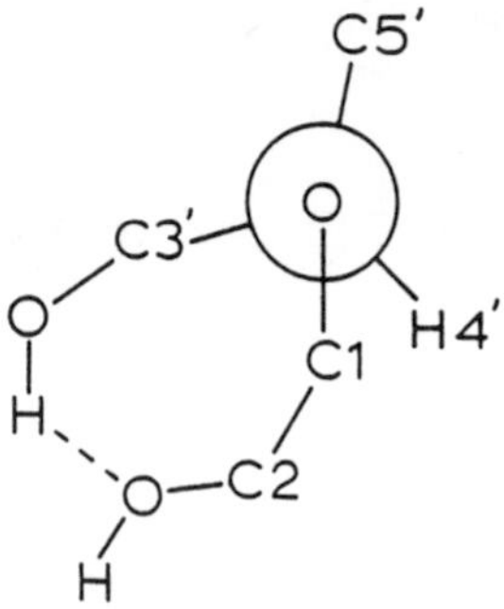


10

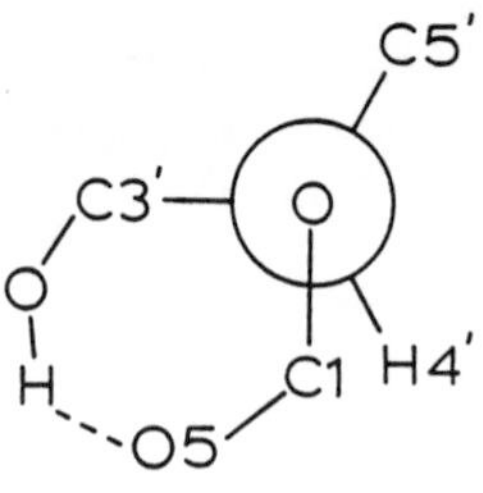


12

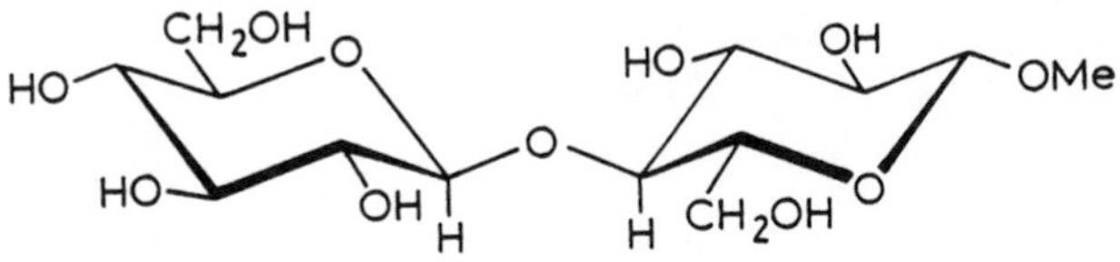


11

$^1\underline{J}_{C-H}$ is related in a consistent fashion to the orientation of
the C-1,H-1 bond: for an equatorial bond (as in 6a), the coupling
is about 10 Hz larger than when, in the anomer (6b), the C-1,H-1
bond is axial (e.g., 170 Hz vs 160 Hz) (12,22,41-44). Also important is the fact that the signal of an anomeric carbon is usually
well separated from other signals in the spectrum, as already
noted. Hence, problems of configuration and conformation may be
examined in a facile manner by $^{13}C$-NMR.

Stereochemistry of $^{13}C$-$^1H$ coupling in oligosaccharide models.

With monosaccharides, $^1H$-coupled spectra show splitting of
up to ∿8 Hz due to two- and three-bond coupling ($^2J$ and $^3J$), both
of which are stereochemically dependent (12,41,45-47). These fine
splittings are not generally resolved in polysaccharide spectra.
If, however, one considers an oligosaccharide as a model for polysaccharide conformation, the observed $^{13}C$-$^1H$ coupling across the
glycosidic bond ($^3J_{COCH}$) in the smaller molecule (e.g. 7) can
provide information about the magnitude of inter-residue torsional
angles ($\phi$ and $\psi$) characteristic of that particular type of linkage (48). This requires a knowledge of the angular dependence of
$^3J$. Although data are available from carbohydrate model compounds
for a variety of $^{13}C$-O-C-$^1H$ arrays of nuclei, it is difficult to
find molecules that can furnish appropriate $^3J$ values for the 0°-
60° region of the Karplus curve relating $^3J$ to the dihedral
angle. One suitable molecule is cyclohexaamylose (formula 8 depicts three of its six glucose residues). According to crystallo-
graphic and theoretical studies the glycosidic bonds are so con-
strained by its cyclic structure that $\phi$(the angle relating H-1
with C-4 in 8) and $\psi$(that relating C-1 with H-4) are about ±10°.

Although C-1 (or C-4) in 8 may couple with several protons
other than H-4 (or H-1), and hence give rise to a complex signal
(49), the introduction of deuterium at positions 2,3 and 6
through facile catalytic H-D exchange (50) improves things materially. Analysis of the C-1 signal for deuterated 8, with the aid
of computer simulation has given a value of 4.8 Hz due to the
inter-residue coupling between C-1 and H-4, and of 5.2 Hz due to
that between C-4 and H-1. With these data, the Karplus curve has
been extended to cover the overall range, as shown in Fig. 6.

This curve could then be used in examining the conformations
of disaccharide models of interest. Methyl β-maltoside (9), for
which inter-residue couplings of 2.5-3.0 Hz have been estimated
(49), may be depicted by orientations of the linkage region in
which the time averaged torsional angles are of the order of 45-
50° (10) (the alternative of ∿140° is regarded as improbable on
the basis of other data for 9 from experimental and theoretical
sources). Hence torsional angles $\phi$ and $\psi$ of the β-maltose moiety
in aqueous solution are much larger than those -- 0°-15° -- that
characterize the molecule in the solid state.

Measurements of inter-residue coupling in methyl

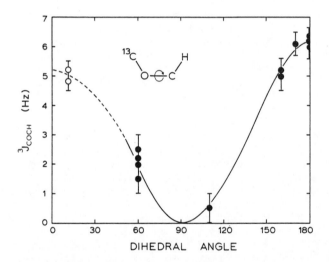

*Figure 6.    A plot of $^3J_{C\text{-}H}$ vs. torsional (dihedral) angle for $^{13}C\text{-}O\text{-}C\text{-}^1H$ arrays of nuclei: (○), values obtained from the spectrum of cyclohexaamylose-$d_4$.*

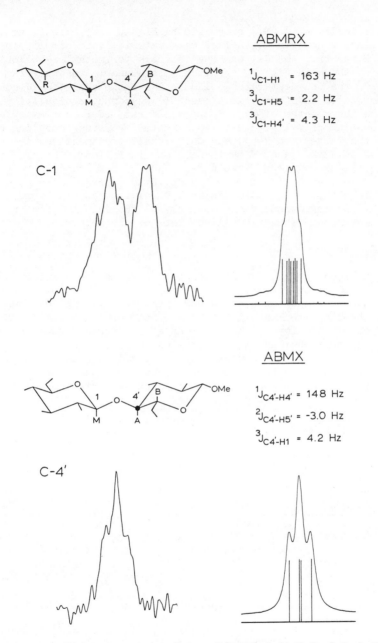

ABMRX

$^1J_{C1\text{-}H1}$ = 163 Hz

$^3J_{C1\text{-}H5}$ = 2.2 Hz

$^3J_{C1\text{-}H4'}$ = 4.3 Hz

C-1

ABMX

$^1J_{C4'\text{-}H4'}$ = 148 Hz

$^2J_{C4'\text{-}H5'}$ = -3.0 Hz

$^3J_{C4'\text{-}H1}$ = 4.2 Hz

C-4'

*Figure 7.   (Upper) C-1 signal of methyl β-cellobioside-d₈ (in D₂O solution). The multiplet to the left that overlaps it partially is the C-1' signal. Computer-simulated C-1 signal is shown on the right. (Lower left), C-4' signal of methyl β-cellobioside-d₈; and (lower right), computer-simulated C-4' signal.*

β-cellobioside (11) (51) proved to be highly complex but, as with
8, extensive deuteration helped very materially in simplifying
the $^{13}$C spectra. The C-1 and C-4' signals for methyl β-cellobio-
side-d$_8$, shown in Fig. 7, are much more amenable to analysis than
those of the non-deuterated molecule. Nevertheless, the $^3J$ coupling constants have been extracted (51) by computer simulation,
because the spectra of Fig. 7 are not first order. For example,
the coupling of C-1 with H-4' is affected by strong coupling between the latter and H-5', and the overall treatment required is
that of an ABMRX case. With the aid of simulation, the value of
inter-residue coupling between C-1 and H-4' was found to be 4.3
Hz, and between C-4' and H-1, 4.2 Hz. Hence, torsional angles φ
and ψ in methyl β-cellobioside in D$_2$O solution are ∿30° (12),
values that correspond rather closely to those in the conformation favored in the crystalline state of the molecule.

Summary.

In recent years, $^{13}$C-NMR spectroscopy has found extensive
use in studies on carbohydrate polymers, in some series over-
shadowing the importance of $^1$H-NMR spectroscopy. Applications
range through determination of primary structure, analysis of
mixtures, monitoring of chemical and enzymic transformations or
of chelation reactions, and studies on conformational change.
Measurements of $^{13}$C-$^1$H coupling are utilized in determining the
configuration and conformation of glycosidic bonds and, in addition, difference methods and simulation experiments are employed
as an aid in the analysis of complex spectra .

Acknowledgments.

The authors express their gratitude to the National Research
Council of Canada and the Faculty of Graduate Studies, McGill
University, for generous support.

Literature Cited.

1.  W.M. Pasika and L.H. Cragg, Can. J. Chem., 41, 293,777(1963).
2.  A.S. Perlin, M. Mazurek, L.B. Jaques and L.W. Kavanagh,Carbo-
    hyd. Res., 7, 369(1968); L.B. Jaques, L.W. Kavanagh, M. Maz-
    urek and A.S. Perlin, Biochem. Biophys. Res. Commun., 24,
    447(1966).
3.  A.S. Perlin, B. Casu, G.R. Sanderson and L.F. Johnson, Can.
    J. Chem., 48, 2260(1970).
4.  A. Darke, E.G. Finer, R. Moorhouse and D.A. Rees, J. Mol.
    Biol., 99, 477(1975).
5.  M.W. Duch and D.M. Grant, Macromolecules, 3, 165(1970).
6.  J. Schaeffer, Macromolecules, 4, 110(1971).
7.  H. Saito, T. Ohki and T. Sasaki, Biochemistry, 16, 908
    (1977).

8.  H. Saito, T. Ohki, N. Takasuka and T. Sasaki, Carbohyd. Res., 58, 293(1977).
9.  D.E. Dorman and J.D. Roberts, J. Am. Chem. Soc., 93, 4463 (1971).
10. A.S. Perlin, N.M.K. Ng Ying Kin, S.S. Bhattacharjee and L.F. Johnson, Can. J. Chem., 59, 2437(1972).
11. A.S. Perlin, in G.O. Aspinall (Ed.), M.T.P. Rev. Sci., Org. Chem., Series 2, Carbohydrates, 7, 21(1976).
12. A.S. Perlin and B. Casu, Tetrahedron Lett., 2921(1969).
13. L.D. Hall and L.F. Johnson, Chem. Commun., 509(1969).
14. D.E. Dorman and J.D. Roberts, J. Am. Chem. Soc., 92, 1355 (1970).
15. A.S. Perlin, B. Casu and H.J. Koch, Can. J. Chem., 48, 2596 (1970).
16. H.J. Jennings and I.C.P. Smith, J. Am. Chem. Soc., 95, 606 (1973).
17. P.A.J. Gorin, Carbohyd. Res., 39, 3(1975).
18. A. Parfondry and A.S. Perlin, Carbohyd. Res., 57, 39(1977).
19. A.S. Perlin, International Symposium on Macromolecules (IUPAC), Rio de Janeiro, Brazil, July 26, 1974.
20. A.S. Perlin, Federation Proceedings, 36, 106(1977).
21. E. Mushayakarara, M.Sc. thesis, McGill University, Montreal, Quebec, August 1977.
22. G.K. Hamer and A.S. Perlin, Carbohyd. Res., 49, 37(1976).
23. M. Vincendon, Bull. Soc. Chim. Fr., 3501(1973).
24. P.A.J. Gorin and J.F.T. Spencer, Can. J. Microbiol., 18, 1709(1972).
25. P.A.J. Gorin, Can. J. Chem., 51, 2105(1973).
26. T, Usui, N. Yamaoka, K. Matsuda, K. Tuzimura, H. Sugiyama, S. Seto, J. Chem. Soc. Perkin Trans. I, 2425(1973).
27. D. Gagnier, R. Marchessault, M. Vincendon, Tet. Lett.53(1975)
28. G.K. Hamer, S.S. Bhattacharjee and W. Yaphe, Carbohyd. Res., 54, C7(1977).
29. A.K. Bhattacharjee, H.J. Jennings, C.P. Kenny, A. Martin and I.C.P. Smith, J. Biol. Chem., 250, 1926(1975).
30. D.R. Bundle, I.C.P. Smith and H.J. Jennings, J. Biol. Chem., 249, 2275(1974).
31. G.G.S. Dutton, K.L. Mackie, A. V. Savage and M.D. Stephenson, Carbohyd. Res., 66, 125(1978).
32. S.S. Bhattacharjee, W. Yaphe and G.K. Hamer, IX th International Seaweed Symposium, Santa Barbara, California (1977).
33. A.S. Shashkov, A.I. Usov and S.V. Yarotsky, Bioorgan. Khimiya, 4, 74(1978).
34. S.V. Yarotsky, A.S. Shashkov and A.I. Usov, Bioorgan. Khimiya, 3, 1135(1977).
35. S.S. Bhattacharjee, W. Yaphe and G.K. Hamer, Carbohydr. Res., 60, C1(1978).
36. G.K. Hamer, unpublished results.
37. J.W. Blunt and M.H.G. Munro, Aust. J. Chem., 29, 975(1976).
38. P.A.J. Gorin and M. Mazurek, Can. J. Chem., 51, 3277(1973).

39.  G. Gatti, B. Casu, N. Cyr and A.S. Perlin, Carbohyd. Res.,
     41, C6(1975).
40.  A.S. Perlin, G.K. Hamer, B. Casu and G. Gatti, Polymer Pre-
     prints, 19, No. 2, 10(1978).
41.  J.A. Schwarcz and A.S. Perlin, Can. J. Chem., 50, 3667(1972).
42.  K. Bock, J. Lundt and C. Pederson, Tetrahedron Lett., 1037
     (1933).
43.  K. Bock and C. Pederson, J. Chem. Soc. Perkin Trans. II,
     293(1974).
44.  F.R. Taravel and P.J.A. Vottero, Tetrahedron Lett. 2341
     (1975).
45.  J.A. Schwarcz, N. Cyr and A.S. Perlin, Can. J. Chem., 53,
     1872(1975).
46.  N. Cyr, G.K. Hamer and A.S. Perlin, Can. J. Chem., 56, 297
     (1978).
47.  R.U. Lemieux, T.L. Nagabhushan and B. Paul, Can. J. Chem.,
     50, 773(1972).
48.  A.S. Perlin, N. Cyr, R.G.S. Ritchie and A. Parfondry, Carbo-
     hyd. Res., 37, C1(1974).
49.  A. Parfondry, N. Cyr and A.S. Perlin, Carbohyd. Res., 59,
     299(1977).
50.  F. Balza, N. Cyr, G.K. Hamer, A.S. Perlin, H.J. Koch and
     R.S. Stuart, Carbohyd. Res., 59, C7(1977); H.J. Koch and
     R.S. Stuart, Carbohyd. Res., 59, C4(1977).
51.  G.K. Hamer, F. Balza, N. Cyr and A.S. Perlin, Can. J. Chem.,
     in press.

Discussion.

   R.H. Marchessault (Xerox, Mississauga): You are to be con-
gratulated on carrying out the analysis of the dihedral angles,
because I know this is a difficult thing to do and that you have
been trying to do it for some time in Dr. Perlin's lab. The pre-
cision you got though, I would guess, is in the neighborhood of
± 20 degrees, indicating the reliability of the measurement.
Unequivocally one can say you are on one side of the curve or the
other. For the cellobiose case the results seem to fall in with
what we expect for β-glycosides. Can the reliability of the
measurement be increased? Obviously, the crystallographer has an
ideal case. He is dealing with a single crystal where everything
is stable and there is only one angle involved. There must be a
certain amount of averaging in your case.
   G.K. Hamer: That is certainly true. I think the problem is
not that we can't measure the coupling constants. I think the
accuracy with which we can measure the coupling constants is
pretty good because it turns out that the simulations are extremely sensitive to small changes in chemical shifts, and the
changes in chemical shifts that are important are in the proton
spectrum. They are minute changes. The real problem is in the
Karplus equation and in parameterising the curve. I think one

can't push it a great deal further. Another point involves the coupling pathways. One is H-1 through to C-4', the other is H-4' through to C-1. They are not strictly identical because in the one case C-1 is attached to two oxygens whereas C-4' is attached to only one oxygen. The question is, can one use the same curve to derive angles for both interglycosidic bonds?

R.H. Marchessault: I would like to encourage you to keep going because its the first direct measure of the $\phi$-$\psi$ angles in solution that we have seen.

F. Seymour (Baylor Medical Center, Texas): In the heparin spectrum for the hog mucosa there was a rather prominent peak around 25 p.p.m. Do you know its origin? Can you comment on that peak, and its relation to the spectrum of normal bovine lung?

G.K. Hamer: Yes, that peak comes from the N-acetyl function which is substituted for N-sulfate on the glucosamine.

F. Seymour: Is it present in the spectrum of the bovine lung?

G.K. Hamer: No, it is not. Using the educated eye (once you know its there in the A type) if you look at the B type very closely you see a tiny peak in the same position. If you look at the 220 MHz proton spectrum it is much more obvious. There is a very small amount in the bovine. The main differences between the two sources are a) N-acetyl substitution for N-sulfate and b) the change in uronic acid.

RECEIVED March 13, 1979.

# Relaxation Studies in the System Poly(ethyl methacrylate)–Chloroform by Carbon-13 and Proton NMR

LENAS J. HEDLUND, ROBERT M. RIDDLE, and WILMER G. MILLER

Department of Chemistry, University of Minnesota, Minneapolis, MN 55455

Nuclear magnetic resonance spectroscopy of dilute polymer solutions is utilized routinely for analysis of tacticity, of copolymer sequence distribution, and of polymerization mechanisms. The dynamics of polymer motion in dilute solution has been investigated also by proton[1-7] and by carbon-13[5-15] NMR spectroscopy. To a lesser extent the solvent dynamics in the presence of polymer has been studied.[16-18] Little systematic work has been carried out on the dynamics of both solvent and polymer in the same system.

The concentration dependence of polymer or solvent motion has been studied only rarely over a wide range in concentration.[19-22] Typically, polymer carbon-13 relaxation is not concentration dependent up to 20-30 percent polymer. Little is known concerning the concentration dependence of the solvent motion.

Carbon-13 relaxation depends predominantly on intramolecular contributions, whereas proton relaxation is sensitive to intermolecular as well as intramolecular interactions. However, by use of isotope dilution[23] the two types of interactions may be separated. Studies utilizing both nuclei can thus yield complimentary information.

The experimental design was to study both the carbon-13 and proton relaxation as a function of temperature for both polymer and solvent, and to extend these to as high a polymer concentration as the available equipment permitted. Inasmuch as the mechanical properties of polymers can be affected considerably by small amounts of diluents, we would ultimately like to approach the bulk polymer state, where use of strong dipolar decoupling and magic angle spinning are necessary.[24]

We chose the system poly(ethyl methacrylate)-chloroform (PEMA-CHCl$_3$) for several reasons. Karim and Bonner,[25] using a PEMA packed gas chromatography column, have shown that CHCl$_3$ has the strongest interaction with bulk PEMA out of the thirty solvents investigated. Secondly, although PEMA has not been so thoroughly studied as poly(methyl methacrylate), PMMA, its chain dynamics in solution should resemble PMMA, which in bulk has also been studied.[26] Finally, the dynamics of neat chloroform has been studied by NMR[27-32] and by dielectric relaxation.[33,34]

0-8412-0505-1/79/47-103-143$05.00/0
© 1979 American Chemical Society

Experimental

Poly(ethyl methacrylate) (Cellomer Associates) was vacuum
dried at 50°C. The molecular weight ($M_w$) was determined to be
3.3 x 10⁵ from its intrinsic viscosity in ethyl acetate.[35]
Chloroform (spectral grade) and deuterochloroform (MSD Isotopes)
were used as received. Prior to sample preparation the solvent
was degassed using five freeze-thaw cycles. The solvent was
vacuum distilled onto the polymer in a 12 nm NMR tube, and sealed.
¹H and ¹³C spin-lattice relaxation times were made on a
Varian Associates XL-100 - 15/VFT-100 Spectrometer operating at
100.1 and 25.2 MHz, respectively, using differential mode[36] inversion recovery or saturation recovery[37] using homospoil. ¹H
and ¹³C $T_1$ measurements were made using either an external ¹⁹F
or an internal ²H field frequency lock. Either integrated peak
areas or peak intensities were used to determine $T_1$ from a linear
plot of intensity versus delay depending on the linewidth of the
resonance. For sharp resonances no difference was found in $T_1$'s
determined from integrated areas or intensities.

Results

The ¹³C spectrum of 6, 20 and 40 wt. % PEMA solutions at
34°C are shown in Figure 1. All of the resonances are easily
discernible except for the backbone methylene at 40%. At low
concentration the polymer $\alpha$-CH₃, quaternary carbon, and backbone
methylene carbon exhibit resolved or partially resolved chemical
shifts due to the various stereochemical sequences since the
polymer was not stereoregular. A rough estimate indicates the
polymer is essentially atactic.
The ¹³C relaxation behavior of chloroform as a function of
temperature and polymer concentration is shown in Figure 2, and
the corresponding nuclear Overhauser enhancement factors in
Figure 3. The values for the neat solvent are in rather good
agreement with literature values.[28] Addition of as little as
three percent polymer is seen to have a measurable affect on the
solvent relaxation. As the polymer concentration is increased
further there is a systematic lowering of $T_1$. At 30% and higher,
a $T_1$ minimum is observed.
In Figures 4-6 the temperature and concentration dependence
of the quaternary carbon, $\alpha$ and ester methyls, and ester methylene
¹³C relaxations are shown. Data on the backbone methylene have
not yet been accumulated, nor have the nuclear Overhauser enhancements on the polymer carbons. At low polymer concentration
the difference in $T_1$ among the stereotriads for a given ¹³C was
less than experimental error. At higher concentration only an
average could be determined. In general the concentration dependence of $T_1$ is considerably less than that observed with the
solvent. A $T_1$ minimum is found for the ester methylene, and the
quaternary carbon relaxation. The minimum moves to higher temperature with increasing polymer concentration, analogous to the

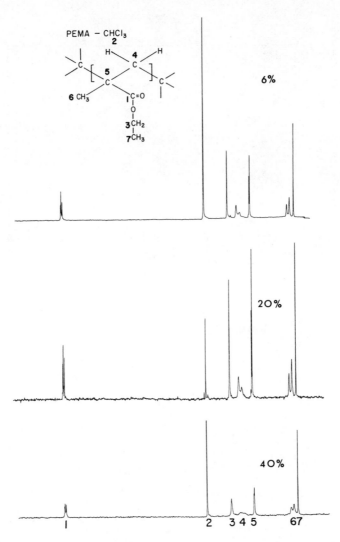

Figure 1.   *C-13 spectrum of 6, 20, and 40 wt % PEMA solutions at 34°C*

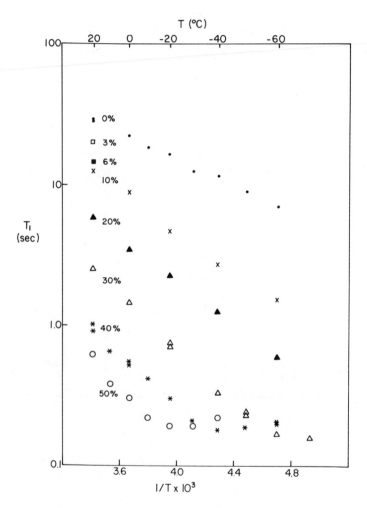

*Figure 2.   Chloroform C-13 relaxation as a function of temperature. The polymer concentration (weight percent) is as indicated.*

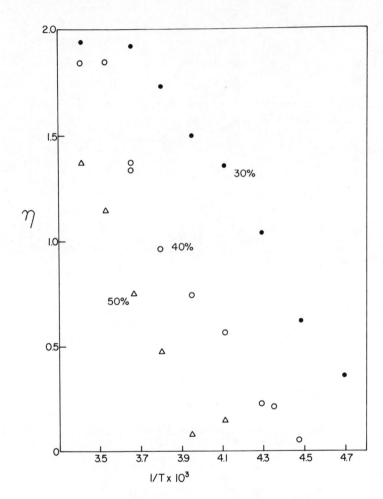

*Figure 3. Chloroform nuclear Overhauser enhancement factor as a function of temperature at the indicated polymer concentrations.*

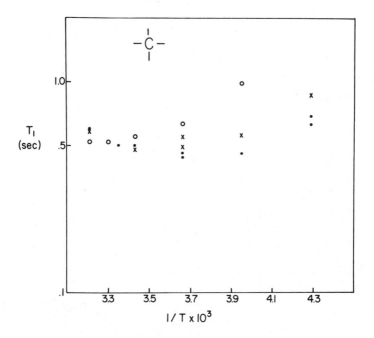

*Figure 4.   Temperature and concentration dependence of the quaternary C-13
relaxation: (●), 10%; (×), 20%; (○), 30%.*

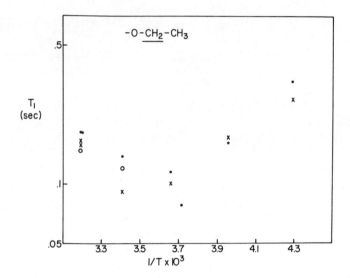

*Figure 5. Temperature and concentration dependence of the ester methylene C-13 relaxation: (●), 10%; (×), 20%; (○), 30%.*

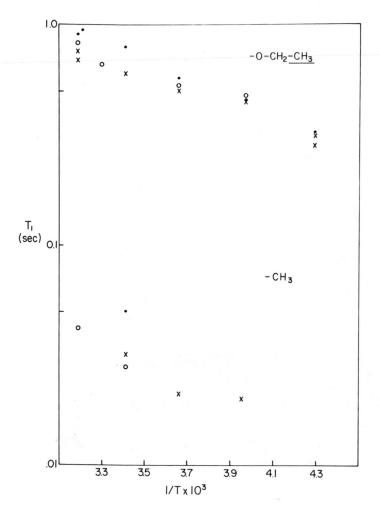

Figure 6.  Temperature and concentration dependence of the methyl C-13 relaxation: ($\bullet$), 10%; ($\times$), 20%; ($\bigcirc$), 30%.

results for the solvent. The ester methyl has a considerably
larger $T_1$ than the ester methylene due to the internal rotation
of the methyl. The $\alpha\text{–CH}_3$ $T_1$ is more than an order of magnitude
smaller than that of the ester methyl, as it reflects more closely
the backbone motion and probably a higher barrier for internal
rotation as well.[38]

In Figure 7 is shown the solvent [1]H relaxation time as a
function of temperature and polymer concentration. Analogous to
the [13]C results, the proton relaxation is affected by addition
of only a small amount of polymer. Further addition of polymer
decreases $T_1$ systematically.

The neat chloroform proton $T_1$ data are in substantial agreement with those of Bender and Zeidler.[27] Unlike chloroform [13]C
relaxation, which is almost exclusively by intramolecular dipole-
dipole relaxation, [1]H relaxation has intermolecular as well as
intramolecular contributions. By means of dilution with $CDCl_3$
one can obtain separately the intramolecular and the various
intermolecular contributions.[27] Additional intermolecular terms
must be included in the presence of polymer. Expressed in terms
of relaxation rates,

$$(1/T_1) = (1/T_1)_{intra} + (1/T_1)_{inter} \tag{1}$$

where

$$(1/T_1)_{intra} = (1/T_1)_{intra}^{spin\ rotation} + (1/T_1)_{intra}^{H-Cl} \tag{2}$$

and

$$(1/T_1)_{inter} = (1/T_1)_{inter}^{H-Cl} + (1/T_1)_{inter}^{H-HS} + (1/T_1)_{inter}^{H-P}. \tag{3}$$

The superscripts H–Cl refers to solvent proton-chlorine interaction, H–HS to solvent proton-proton interaction, and H–P to
interaction of the solvent proton with the polymer. The latter
should be dominated by solvent proton-polymer proton interaction.
Upon addition of $CDCl_3$, followed by extrapolation to pure $CDCl_3$,
the extrapolated rate $[(1/T_1)_0]$ is given by

$$(1/T_1)_0 = (1/T_1)_{intra} + (1/T_1)_{inter}^{H-Cl} + (1/T_1)_{inter}^{H-D} + (1/T_1)_{inter}^{H-P} \tag{4}$$

Equations 1–4, plus a knowledge[39] of the theoretical ratios
$(1/T_1)_{inter}^{H-D} \Big/ (1/T_1)_{inter}^{H-H}$ and $(1/T_1)_{inter}^{H-Cl} \Big/ (1/T_1)_{inter}^{H-H}$, allows the

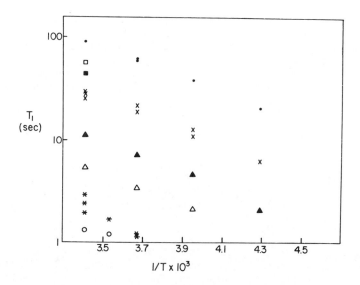

*Figure 7. Temperature and concentration dependence of the chloroform ¹H relaxation:* (●), 0%; (□), 3%; (■), 6%; (×), 10%; (▲), 20%; (△), 30%; (*), 40%; (○), 50%.

intramolecular and intermolecular contributions to $T_1$ to be assessed. In pure chloroform the intramolecular relaxation is found to be dominated by spin rotation.[27] The intermolecular contributions dominate the intramolecular at all temperatures, with the proton–proton interaction being the major contributor to the intermolecular relaxation mechanisms. In the presence of polymer the quantity $(1/T_1)_{inter}^{H-P} + (1/T_1)_{intra}$ is more easily determined from the experimental data than the proton–polymer interaction alone. Shown in Figure 8 is the solvent proton–solvent proton contribution as a function of polymer concentration. Shown in Figure 9 is the solvent proton–polymer contribution, where we have assumed the intramolecular contribution is small, and also scaled the results to reflect the number of polymer protons present.

## Discussion

In a polymer solution the translational diffusion of solvent is well known to be related to the internal motion of the polymer, and its concentration dependence.[40-42] It would not be surprising to find a relationship between solvent rotational motion and polymer segmental motion, particularly at high polymer concentration. There is little systematic literature bearing on this point. Rothschild,[43] by analyzing the band shape of a far infrared band of $CH_2Cl_2$, found only a small change in solvent rotational correlation time upon addition of up to 60% polystyrene. Anderson and Liu,[16] studying the proton relaxation of benzene in the presence of PMMA, found a systematic decrease in $T_1$ with increasing polymer concentration. Addition of 35% PMMA resulted in a factor of three reduction in $T_1$. Through the use of isotope dilution they were able to separate the relaxation into intramolecular and intermolecular contributions. Inasmuch as their analysis indicated that the intramolecular relaxation was independent of polymer concentration, they concluded that up to 35% PMMA had no effect on the rotational motion of benzene. Finally we turn to the work of Heatley and Scrivens,[18] who measured the $^{13}C$ and $^1H$ relaxation in acetone at 0, 5, 10 and 20% PMMA. At high temperatures spin rotation dominated the relaxation and little information was obtainable concerning solvent motion. At low temperatures, where dipolar relaxation dominated, no effect of polymer on acetone rotation was evident up to 10% PMMA, but quite perceptible at 20% PMMA.

Returning to our data, it is especially interesting that we find a systematic reduction in solvent $^{13}C$ relaxation upon increase in polymer concentration, even at low polymer concentration, inasmuch as the relaxation should be almost exclusively by intramolecular dipolar relaxation.[28] The existence of a $T_1$ minimum at higher polymer concentration permits us to make a meaningful comparison with various models. If the chloroform $^{13}C$ relaxation

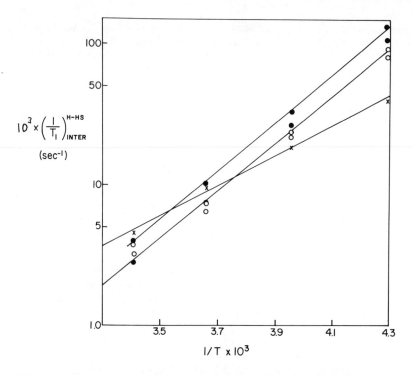

*Figure 8. Temperature and concentration dependence of the solvent proton–solvent proton relaxation rate: (●), 20%; (○), 10%; (×), 0%.*

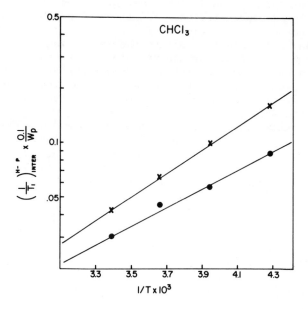

*Figure 9. Temperature and concentration dependence of the solvent proton–polymer proton relaxation rate: (●), 10% polymer; (×), 20% polymer.*

was describable by a single isotropic rotational correlation time, $\tau_c$, the spectral densities $J_n(\omega)$ would be given by

$$J_n(\omega) = \tau_c / (1+\omega^2\tau_c^2), \qquad (5)$$

leading to a $T_1$ minimum of 0.040 sec. The fact that the $T_1$ minimum is more than 4 times larger in itself rules out a single rotational correlation time. The next logical step is to assume a distribution of correlation times, as it will raise the $T_1$ minimum. The appropriate spectral densities for both symmetric and asymmetric distributions of isotropic rotations have been given[10,44], wherein the relaxations are characterized by a mean correlation time and a distribution width parameter. Keeping the width parameter and type of distribution function fixed, $T_1$ as a function of mean correlation time may be calculated. Assuming, for instance, the symmetric Cole–Cole distribution, a width parameter of 0.25 will give a $T_1$ minimum in agreement with the 30% PEMA data. However, upon keeping the width parameter constant, a mean correlation time of less than $10^{-15}$ sec. would be required to yield a calculated $T_1$ in agreement with the $20^\circ$C date. This is clearly a meaningless value as solvent rotational correlation times are never that small. As an example, for pure deuterochloroform at $20^\circ$, Huntress[29] calculates $\tau_\perp$ to be 1.8 x $10^{-12}$ sec. and $\tau_{\parallel}$ to be 0.92 x $10^{-12}$ sec. It is highly unlikely that the rotational diffusion of the solvent in a polymer solution could possibly exceed that of the pure solvent. Additionally the calculated $\eta$ corresponding to this mean $\tau$ and distribution width is much lower than the experimental value at $20^\circ$C of nearly 2. Unsymmetric distributions give unsatisfactory results also. The assumption of isotropic motion is not a significant source of error, as chloroform rotation diffusion coefficients differ by less than a factor of two.[29] Relaxation by scalar coupling and spin rotation is small in pure chlorform[28], and we can see little reason why it should be more significant in the presence of polymer. Hence another mechanism must be devised to explain the experimental results.

A possible step in this direction can be made through use of earlier relaxation studies on other systems. Hunt and Powles,[45] when studying the proton relaxation in liquids and glasses, found the relaxation best described by a "defect-diffusion" model, in which a non-exponential correlation function corresponding to diffusion is included together with the usual exponential function corresponding to rotational motion. The correlation function is taken as the product of the two independent reorientation processes. This type of model has more recently been applied to $^{13}$C relaxation in polymers,[10] where the nonexponential part corresponds to conformational jumps in a Monnerie type[46] local mode

mechanism plus an exponential contribution corresponding to overall tumbling. A continium version of this has recently been
given.[47] This can at least qualitatively explain the experimental
data in that the $T_1$ minimum can be significantly increased over a
single exponential model while keeping the nuclear Overhauser
enhancement factor much closer to the experimental values, while
at the same time not leading to physically absurd values for the
exponential reorientation. One can give a physical picture to
this scheme which makes some intuitive sense. In the presence of
polymer the solvent is relaxed by two mechanisms. One is the
rotation of the solvent, an exponential reorientation contribution. The nonexponential or diffusive contribution could result
from the reorientation when the solvent drops into the hole resulting from the conformational jump of the polymer. This would
be compatible with the mechanisms used to explain the relationship between solvent translational diffusion and polymer segment
motion, as mentioned earlier. The solvent proton results can
also be analyzed in terms of this mechanism, as well as the polymer backbone motion. We are currently trying to quantitatively
analyze the data from this viewpoint.

References

1.  K. J. Liu and R. Ullman, J. Chem. Phys., 48, 1158 (1968).

2.  K. J. Liu and W. Burlant, J. Polym. Sci., Part A-1, 5, 1407
    (1967).
3.  K. Hatada, Y. Okamoto, K. Ohta, and H. Yuki, J. Polym. Sci.,
    Polym. Letters Ed., 14, 51 (1976).

4.  F. Heatley and M. K. Cox, Polymer, 18, 225 (1977).

5.  D. Ghesquiere, B. Ban, and C. Chachaty, Macromolecules, 10,
    743 (1977).

6.  J. Spevacek and B. Schneider, Polymer, 19, 63 (1978).

7.  D. Doskocilova, B. Schneider, and J. Jakes, J. Mag. Res.,
    29, 79 (1978).

8.  J. Schaefer, Macromolecules, 6, 882 (1973).

9.  J. R. Lyerla and T. T. Horikawa, J. Polym. Sci., Polym.
    Letters Ed., 14, 641 (1976).

10. F. Heatley and A. Begum, Polymer, 17, 399 (1976).

11. F. A. Bovey, F. C. Schilling, T. K. Kwei and H. L. Frisch,
    Macromolecules, 10, 559 (1977).

12. K. Hatada, T. Kitayama, Y. Okamoto, K. Ohta, Y. Umemura and
    H. Yuki, Makromol. Chem., 178, 617 (1977).

13. R. E. Cais and F. A. Bovey, Macromolecules, 10, 752 (1977).

14. R. E. Cais and F. A. Bovey, Macromolecules, 10, 757 (1977).

15. W. H. Stockmayer, A. A. Jones and T. L. Treadwell, Macro-molecules, 10, 762 (1977).

16. J. E. Anderson and K. J. Liu, J. Chem. Phys., 49, 2850 (1968).

17. K. Sato and A. Nishioka, J. Polym. Sci. Part A-2, 10, 489 (1972).

18. F. Heatley and J. H. Scrivens, Polymer, 16, 489 (1975).

19. W. P. Slichter and D. D. Davis, Macromolecules, 1, 47 (1968).

20. F. Heatley, Polymer, 16, 493 (1975).

21. T. Cosgrove and R. F. Warren, Polymer, 18, 255 (1977).

22. R. Kimmich and Kh. Schmauder, Polymer, 18, 239 (1977).

23. G. Bowers and A. Rigamonti, J. Chem. Phys., 42, 171 (1965).

24. J. Schaefer, in "Structural Studies of Macromolecules by Spectroscopic Methods", K. J. Irvin, Ed., John Wiley & Sons, 1976, pp 201-226.

25. A. K. Karim and D. C. Bonner, Soc. Plast. Eng., Tech. Pap., 22, 3 (1976).

26. D. W. McCall, Natl. Bur. Stand. (U.S.) Spec. Publ. No. 301, 475 (1969).

27. H. J. Bender and M. D. Zeidler, Ber. Bunsenges Physik. Chem., 75, 236 (1971).

28. R. Shoup and T. Farrar, J. Mag. Res., 7, 48 (1972).

29. W. T. Huntress, J. Phys. Chem., 73, 103 (1969).

30. D. E. O'Reilly and G. E. Schacher, J. Chem. Phys., 39, 1768 (1963).

31. G. Bowers and A. Rigamonti, J. Chem. Phys., 42, 175 (1965).

32. K. Gillen, M. Schwartz and J. Noggle, Mol. Phys., 20, 899 (1971).

33. S. Mallikarjun and N. E. Hill, Trans. Faraday Soc., 61, 1389 (1965).

34. S. Mallikarjun and N. E. Hill, Rheologica Acta, 5, 232 (1966).

35. J. Brandrup and H. Immergut, Polymer Handbook, 2nd ed., Interscience, New York, 1975.

36. R. Freeman and H. D. W. Hill, J. Chem. Phys., 54, 3367 (1971).

37. J. L. Markley, W. J. Horsley, and M. P. Klein, J. Chem. Phys., 55, 3604 (1971).

38. J. R. Lyerla, T. T. Horikawa and D. E. Johnson, J. Am. Chem. Soc., 99, 2466 (1977).

39.  A. Abragam, The Principles of Nuclear Magnetism, Oxford (1961).

40.  R. S. Moore and J. D. Ferry, J. Phys. Chem., 66, 2699 (1962).

41.  C. P. Wong, J. L. Schrag, and J. D. Ferry, J. Polym. Sci.
     Part A-2, 8, 991 (1970).

42.  J. D. Ferry, "Viscoelastic Properties of Polymers,"
     Wiley, New York (1970).

43.  W. G. Rothschild, Macromolecules, 1, 43 (1968).

44.  J. Schaefer, Macromolecules, 6, 882 (1973).

45.  B. I. Hunt and J. G. Powles, Proc. Phys. Soc., 88, 513 (1966).

46.  B. Valur, J. P. Jarry, F. Geny and L. Monnerie, J. Polym.
     Sci., Polym. Phys. Ed. 13, 667, 675 (1975).

47.  J. T. Bendler and R. Yaris, Macromolecules, 11, 650 (1978).

RECEIVED March 13, 1979.

# The Use of Carbon-13 NMR to Study Binding of Hormones to Model Receptor Membranes: The Opioid Peptide Enkephalin

HAROLD C. JARRELL, P. TANCREDE, ROXANNE DESLAURIERS, and IAN C. P. SMITH

Division of Biological Sciences, National Research Council of Canada, Ottawa, Ontario, Canada K1A 0R6

W. HERBERT McGREGOR

Wyeth Laboratories, P. O. Box 8299, Philadelphia, PA 19101

The recent discovery of a class of peptides, the enkephalins, which act as opiate agonists has led to a number of physical chemical studies aimed at understanding the structure-activity relationships between the enkephalins and the opiates (1-9). The $^1$H spin-spin coupling constants [2-4] of the enkephalins in solution can be interpreted in terms of folded conformations resembling that of morphine in the placement of the residues which appear important for biological activity. X-ray crystallography and theoretical calculations (4-9) have also shown that methionine and leucine enkephalin adopt conformations similar to those concluded from $^1$H NMR studies. Hence it would appear that opioid peptides can topographically resemble the opiates by assuming preferred, folded, conformations. However, earlier studies from this laboratory (10) have shown that $^{13}$C NMR data can be interpreted in terms of a conformationally flexible structure for methionine enkephalin.

Abood and coworkers (11-13) have reported that acidic lipids, in particular phosphatidyl serine (PS), enhanced binding of opiates to membranous components of brain. Furthermore, PS has been shown to bind opiates stereospecifically whereas other acidic lipids bind to a lesser extent and neutral lipids bind very little (11). Loh and coworkers (14-16) have reported that cerebroside

*Issued as N.R.C.C. Publication No. 17068.

sulfate exhibits strong binding of opiates and that the cerebroside sulfate-opiate complex resembles an opiate complex which had
been isolated from brain components. Lipids, therefore, have been
implicated in the interaction of opiates with their receptors.

It is known (17) that opiates and opioid peptides may exhibit
specific receptor binding (high affinity) and nonspecific (low
affinity) binding. Phosphatidyl serine and other lipids exhibit
weak binding of opiates and therefore serve as models for nonspecific binding. Preliminary to a study of high affinity binding
we chose to study the interaction of methionine enkephalin to
various lipids, in particular PS, as a model for binding. In
these studies we have followed a number of $^{13}$C NMR parameters for
the glycine-2 and glycine-3 residues of methionine enkephalin.
The glycyl residues were 90% enriched in $^{13}$C at the $\alpha$-carbons.
This allowed us to measure chemical shifts ($\delta$), line widths ($\Delta\nu$),
and spin-lattice relaxation times ($T_1$) for the $\alpha$-carbons of the
glycine residues as a function of various parameters. It is
generally believed that the binding is electrostatic in nature
(11,15); we therefore sought a pH dependence of the binding of
enkephalin to the lipids. We further sought competition between
morphine and enkephalin for binding to PS.

## Materials and Methods

[2-[2-$^{13}$C]glycine] and [3-[2-$^{13}$C]glycine] methionine
enkephalin (TyrGlyGlyPheMet), enriched to 90% in $^{13}$C, were prepared as described previously (10). Phosphatidyl serine (PS) was
obtained from Serdary Research Laboratories, London, Canada or
Lipid Products, South Nutfield, England. Egg lecithin was obtained from Lipid Products, South Nutfield, England. Cerebroside
sulfate (sulfatides) was obtained from Analabs, Inc. Morphine
sulfate pentahydrate was purchased from Ingram and Bell Limited,
Don Mills (Toronto), Canada. Samples for binding studies were
prepared as described previously (10).

NMR measurements were performed on samples prepared in

deuterium oxide at 30°C.  pH values are meter readings that have
not been corrected for the deuterium isotope effect.  The pH of
samples was varied using hydrochloric acid and sodium hydroxide
diluted in deuterium oxide.

$^{13}C$ NMR measurements were performed on Varian CFT-20 and
XL-100 spectrometers under conditions which have been described
previously (10).  $^{13}C$ chemical shifts are reported with respect
to external tetramethylsilane (TMS).  Dioxane was added to the
samples as an internal standard for linewidth and chemical shift
measurements.

## Results and Discussion

Three parameters are readily obtainable from $^{13}C$ NMR spectra
which may be useful in studying binding interactions:  the chemical shift ($\delta$), the linewidth ($\Delta \nu$) or the apparent or effective
spin-spin relaxation time ($T_2^*$), and the spin-lattice relaxation
time ($T_1$).  $^{13}C$ chemical shifts can reflect steric strain and
change in the electronic environment within a molecule when it
binds to another species.  Spin-lattice and spin-spin relaxation
times can yield information on the lifetimes, sizes and conformations of molecular complexes.

Studies on morphine binding to PS have shown the interaction
to have a maximum at pH 8-9 (11).  In order to interpret the pH-
dependence of binding it is necessary to determine the dissociation constants (pKa) of the titratable groups in each of the
interacting species.  For PS pKa values of 3.7, 4.0 and 7.5 have
been determined for the phosphate, carboxyl, and amino groups,
respectively (18).  Figure 1 shows the $^{13}C$ chemical shift of the
tyrosyl β-carbon as a function of the pH-meter reading of a
solution of methionine enkephalin in deuterium oxide ($D_2O$).  The
pKa (measured in $D_2O$) was determined to be 7.5 for the amino group
of the tyrosine residue while the methionine carboxyl group had a
pK of 3.9.

The pH-dependence of the chemical shifts measured for the

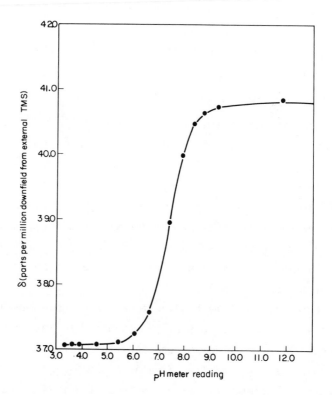

*Figure 1.  Plot of C-13 chemical shift of C-β of tyrosine in enkephalin as a function of pH meter reading in D₂O, 30°C*

enriched gly-2 and gly-3 residues of methionine enkephalin in the absence and presence of PS are shown in Figures 2 and 3, respectively. It is evident that the $^{13}C$ chemical shifts of the glycyl residues were not greatly affected by the presence of lipid. The observed linewidths (corrected for field inhomogeneity) for the gly-2 and gly-3 residues of methionine enkephalin in the presence of PS were also measured as a function of pH; the results are given in Table I. The observed linewidths vary only slightly over the reported pH range. The above results indicate that neither chemical shifts nor linewidths (or $T_2^*$) would be useful for monitoring interaction of enkephalin with PS - at least via the glycyl residues.

Figures 4 and 5 show a pH-dependence of the $T_1$ values of the glycyl residues in [2-[2-$^{13}C$]glycine] and [3-[2-$^{13}C$]glycine] methionine enkephalin (1.1 mg) respectively, in the presence of PS (75 mg). An earlier study (10) showed that the $T_1$ values of the glycyl units of enkephalin free in solution did not change over the pH range 1.5-10.0. In the presence of PS, we observe a dramatic decrease in $T_1$ from that of free enkephalin at "pH" 10, to values close to 60 msec around "pH" 4, for the gly-3 residue. This decrease in $T_1$ values as a function of pH is reversed at "pH" 1.0, where the $T_1$ values again reflect those of free enkephalin. At "pH" values less than 4.0 the intensity of the glycyl resonances was greatly reduced from that of the glycyl resonances of free enkephalin and hence a mapping of the pH-dependence of $T_1$ between "pH" 4.0 and 1.0 was not possible. The above pH-dependence would seem to indicate an ionic-type of interaction between enkephalin and PS. The observed pH-dependence of binding is consistent with interaction between the $NH_3^+$ group of tyrosine in enkephalin and the $CO_2^-$ group of serine in PS, analogous to the earlier model for morphine binding (11). A maximum interaction would be expected when the amino group of tyrosine is fully protonated and the carboxyl group of serine is unprotonated. Thus at pH 10.0 or 1.0 little effect on the $T_1$ values would be expected. It should

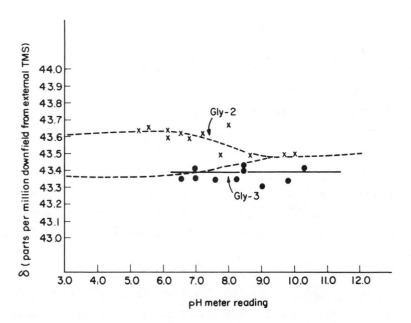

Figure 2.   C-13 chemical shifts of the glycyl residues in [2-[2-C-13]glycine]me-
thionine enkephalin and [3-[2-C-13]glycine]methionine enkephalin in the presence of 75.0 and 78.5 mg of PS, respectively, as a function of pH, 30°C.   Shifts
observed for enkephalin in the absence of PS (see Figure 3), (– – –).

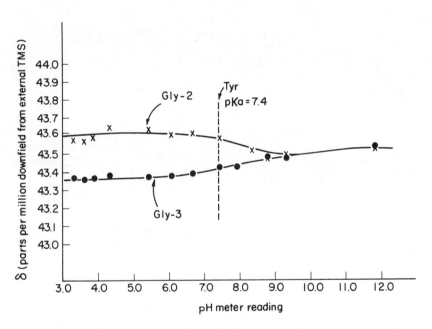

Figure 3. *C-13 chemical shifts of the glycyl-2 and glycyl-3 residues of methionine enkephalin as a function of pH, 30°C*

Table I. Linewidths of $[2-[2-^{13}C]$
glycine] and $[3-[2-^{13}C]$
glycine]Methionine Enkephalin
in the Presence of PS at
Various "pH"'s

| "pH" | $\Delta\nu^a$ (Hz) | "pH" | $\Delta\nu^b$ (Hz) |
|------|------|------|------|
| 5.0 | 5.4 | 6.5 | 5.0 |
| 5.5 | 5.4 | 6.9 | 2.5 |
| 6.1 | 3.2 | 7.6 | 2.7 |
| 6.5 | 3.1 | 8.0 | 3.3 |
| 7.5 | 3.0 | 8.6 | 1.2 |
| 8.1 | 3.0 | 9.8 | 2.0 |

[a] Linewidth for $[2-[2-^{13}C]$glycine] methionine enkephalin (1.1 mg) in the presence of PS (75 mg) in $D_2O$ (1 ml) and corrected for field inhomogeneity.

[b] Linewidth for $[3-[2-^{13}C]$glycine] methionine enkephalin under the same conditions as in (a).

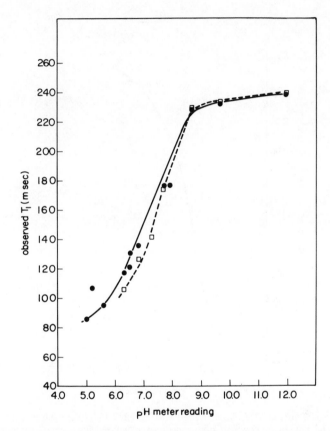

*Figure 4. The pH dependence of T₁ of [2-[2-C-13]glycine]methionine enke-phalin (1.1 mg) in presence of phosphatidyl serine [8.5 mg (□) and 75 mg (●)]. Spectra were run at 20 MHz, 30°C.*

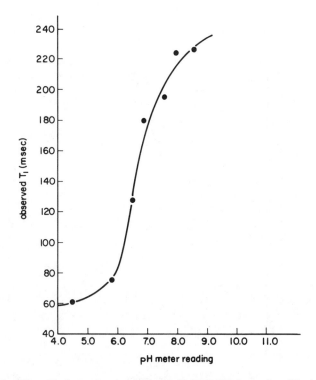

*Figure 5. The pH dependence of* $T_1$ *of [3-[2-C-13]glycine]methionine enkephalin (1.1 mg) in presence of phosphatidyl serine (78.5 mg). Spectra were run at 20 MHz.*

be noted that due to the very similar pKa's of the amino groups of PS and enkephalin, the phosphate and carboxyl group of PS, and the carboxyl group of enkephalin, we cannot determine which groups are specifically involved in binding using the present method.

Rotational correlation times obtained from the observed $^{13}C$ $T_1$ values of the glycyl residues in methionine enkephalin as a function of "pH" are given in Figure 6. The 3-glycyl residue shows a longer correlation time than does the 2-glycyl moiety at low pH. Although binding is thought to occur via the $NH_3^+$ group of the tyrosyl residue, the 2-glycyl unit is known (10) to undergo a greater degree of segmental motion in free solution and this could contribute to the shorter observed correlation time at low pH.

Dissociation Constants. Having established that $T_1$ values could be used to monitor the enkephalin-PS interaction, we sought to obtain a quantitative estimate of the strength of this inter- action. If we assume that we are dealing with a 1:1 peptide-lipid complex, the interaction may be represented as in equation 1,

$$P+L \rightleftharpoons PL \qquad (1)$$

where P and L are methionine enkephalin and lipid, respectively, and PL is the peptide-lipid complex. The dissociation constant $K_d$, which reflects the strength of the interaction, is therefore given by equation 2

$$K_d = \frac{[P_F][L_F]}{[PL]} \qquad (2)$$

where $[P_F]$ and $[L_F]$ are the concentrations of free enkephalin and free lipid, respectively, and $[PL]$ is the concentration of the peptide-lipid complex. In the case of fast exchange it can be shown (19) that the observed relaxation time $T_{1,obs}$ can be related to the mole fraction of the bound species by equation 3

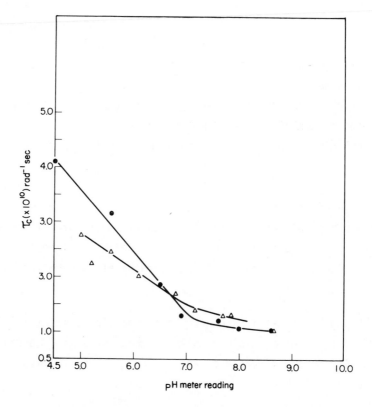

*Figure 6.   Plot of correlation times corresponding to observed* T, *values of [2-[2-C-13]glycine]methionine enkephalin* (×) *and of [3-[2-C-13]glycine]methionine enkephalin* (○) *in presence of PS as a function of pH (1.1 mg of peptide in the presence of ca. 75 mg PS), 30°C*

$$\frac{1}{T_{1,obs}} = \frac{1}{T_{1,F}} + \frac{f_B}{T_{1,B}} \tag{3}$$

where $T_{1,F}$ and $T_{1,B}$ are the spin-lattice relaxation times of the free and bound species and $f_B$ is the mole fraction of the bound species. For an interaction represented by equation 1 it can be shown that the mole fraction bound can be represented by equation 4

$$f_B = \frac{[L_T]}{K_d + [L_T] + [P_T]} \tag{4}$$

where $[L_T]$ and $[P_T]$ are the total concentrations of lipid and enkephalin, respectively. Substituting equation 4 into equation 3 gives:

$$\frac{1}{T_{1,obs}} = \frac{1}{T_{1,F}} + \frac{1}{T_{1,B}} \frac{[L_T]}{K_d + [L_T] + [P_T]} \tag{5}$$

Equation 5 can be rearranged into a more useful form,

$$\frac{T_{1,obs}\, T_{1,F}}{T_{1,F} - T_{1,obs}} = \frac{T_{1,B}\left[K_d + [P_T]\right]}{[L_T]} + T_{1,B} \tag{6}$$

Thus, a plot of $[T_{1,obs}\, T_{1,F}/T_{1,F} - T_{1,obs}]$ vs. $1/[L_T]$ should give $T_{1,B}$ as the intercept and $T_{1,B}(K_d + [P_T])$ as the slope.

We have determined $K_d$ at "pH" 6.3 for $[2-[2-^{13}C]glycine]$ and $[3-[2-^{13}C]glycine]$ methionine enkephalin by varying the lipid concentration in the presence of a 1.2 mg/ml sample of enkephalin; Figures 7 and 8 represent plots of $(T_{1,obs}\, T_{1,F}/T_{1,F} - T_{1,obs})$ vs. $1/[L_T]$. These plots yield $K_d$ values of $5.4 \times 10^{-1}$ M and $3.3 \times 10^{-1}$ M, respectively, and a $T_{1,B}$ value of 35 msec in both cases. An estimate of the sensitivity of the value determined for $T_{1,B}$ and $K_d$ was obtained by varying $T_{1,F}$ by ± 10 msec (within the expected experimental error (10)). This resulted in a change of ± 20 msec for $T_{1,B}$ and a value of $K_d$ which differed by a factor of three.

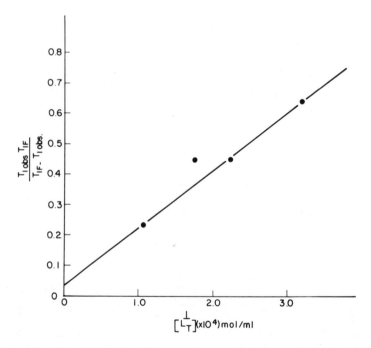

*Figure 7.  Plot of $(T_{1obs}T_{1F})/(T_{1obs} - T_{1F})$ vs. $1/[L_T]$ for [2-[2-C-13]glycine]-methionine  enkephalin.  Spectra were obtained using 1.1 mg of enkephalin at 30°C on a CFT-20 spectrometer operating at 20 MHz.*

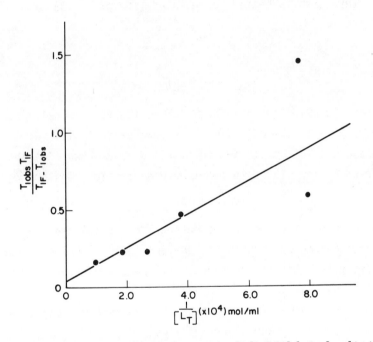

*Figure 8. A plot similar to that in Figure 7 for [3-[2-C-13]glycine]methionine enkephalin*

Using the $T_{1,B}$ obtained above and the pH-dependence of the $T_{1,obs}$ measured for [2-[2-$^{13}$C]glycine] methionine enkephalin (see Figure 4), a pH-dependence of $K_d$ was calculated and is shown in Figure 9. It is readily seen that $K_d$ varies by a factor of five from $3 \times 10^{-1}$ M to 1.6 M. The variation of $K_d$ with pH may be explained by noting that equation 1 assumes that only one ionic form for both enkephalin and PS is involved in binding. However, the interaction is better represented (for pH $\geqslant 6.3$) by

$$P + L \rightleftharpoons PL$$

$$P + L^1 \rightleftharpoons PL^1$$

where $L^1$ is PS in the $NH_2$ form, $PL^1$ is the PS-enkephalin complex involving PS in the $NH_2$ form, P is enkephalin in its zwitterionic form and L is PS as its $NH_3^+$ form. It can be shown (20) that the observed $K_d$ is related to the true dissociation constant $K_d^1$ by

$$K_d = K_d^1 A$$

where A involves the pKa's of enkephalin, PS and the lipid-enkephalin complex. Again the pH-dependence of the observed dissociation constant indicates that the enkephalin-PS interaction is electrostatic in nature.

It is informative to compare the observed dissociation constant for the enkephalin-PS interaction with those reported for other systems. Methionine enkephalin has been shown (21) to bind to brain membrane with a high affinity binding constant of $1.7 \times 10^8$ M$^{-1}$ and a low affinity binding constant of $1.7 \times 10^6$ M$^{-1}$. PS-morphine interaction has been determined to have a $K_d$ of ~$10^{-6}$ M (11). The results observed in this study suggest that the enkephalin-PS binding is weaker than the interaction of morphine with PS.

Correlation Times of the Bound Species. In order to interpret the values of $T_{1,B}$ it is essential to know the rotational correlation time for a particle bound rigidly to a vesicle. An

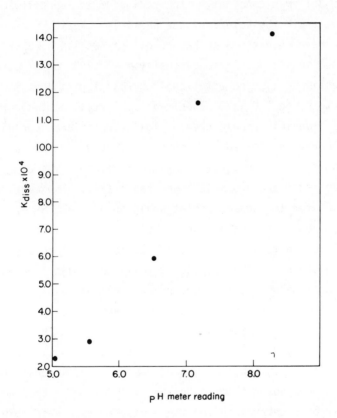

*Figure 9. A plot of $K_d$ vs. pH for [2-[2-C-13]glycine]methionine enkephalin*

estimate of the rotational correlation time for a vesicle can be
determined from the Stokes-Einstein relation

$$\tau = \frac{4\pi r^3 \eta}{3kT} = \frac{V\eta}{kT} \tag{7}$$

where $\eta$ is the viscosity of the solution at a temperature T ($^\circ$K),
k is the Boltzman constant, r the radius of the spherical
particle, and V the volume of the particle. Vesicles of egg
lecithin have been shown to have radii of 105-118 Å (22,23).
Using equation 7 with the viscosity $\eta$ = 0.7808 cP of pure water,
we can estimate $\tau$, the rotational correlation time of the
vesicle to be ca 1.0 x $10^{-6}$ sec rad$^{-1}$. A small particle rigidly
bound to a vesicle should have a similar rotational correlation
time. We have estimated $T_{1,b}$ to be 35 msec at 20 MHz, which
corresponds to possible rotational correlation times of 7.6 x
$10^{-10}$ sec rad$^{-1}$ and 3.0 x $10^{-8}$ sec rad$^{-1}$ (24). The corresponding
linewidths for the bound species would be 9 Hz and 86 Hz. Thus,
it would seem that when enkephalin binds to the lipid vesicle
internal motion (possibly segmental motion) still occurs in the
peptide backbone at the level of glycyl-2 and glycyl-3 residues.
This is not unexpected if the binding involves mainly the $NH_3^+$
group of the tyrosyl unit; this may be termed a one point binding
rather than the possible two point interaction which would involve
both the $NH_3^+$ and $CO_2^-$ group in enkephalin. It would however
appear that the internal motion of the peptide is considerably
decreased on binding from the value of 7.0 x $10^{-11}$ sec rad$^{-1}$
which has been determined for the glycyl-2 residue in solution
(10). This is because fast motions dominate $T_1$ behavior and $\tau_{int}$
values of the order of $10^{-11}$ sec rad$^{-1}$ would result in considerably longer (hundreds of milliseconds) observed $T_{1,B}$ values (23).
The observed $T_{1,B}$ is consistent with a correlation time for
internal motion no smaller than 5.0-7.0 x $10^{-10}$ sec rad$^{-1}$ if the
correlation time of the vesicle is 1.0 x $10^{-6}$ sec rad$^{-1}$ (25).
Thus, it is possible to estimate the rate of internal motion of a

peptide by decreasing the rate of overall motion of the particle
to the extent that the internal motion dominates the spin-
lattice relaxation behaviour.

Linewidths. Observed linewidths can sometimes be useful in
determining which possible value of $\tau$, determined from $T_1$, corresponds to the experimental situation. In the present case, our
$\tau$ values for the bound species should give linewidths of 9 or 86
Hz. Knowing $K_d$, and the linewidth of free enkephalin, we can
calculate an expected "observed" linewidth for all of our experimental situations. Inspection of Table I reveals that no linewidth greater than 7 Hz was observed. As an example of the
linewidth to be expected for conditions of maximal binding to PS,
we choose the case of [2-[2-$^{13}$C]glycine] methionine enkephalin
(1.1 mg) in the presence of PS (75 mg) in $D_2O$ (1 ml) at "pH" 5.0.
The measured $T_1$ was 86 msec and the observed corrected linewidth
was 7 Hz. Equation 8 can be derived

$$\Delta\nu_{obs} = \Delta\nu_F + \Delta\nu_B \frac{[L_T]}{K_d + [L_T] + [P_T]} \qquad (8)$$

in the same manner as equation 5, where $\Delta\nu_{obs}$ is the observed
linewidth, $\Delta\nu_F$ and $\Delta\nu_B$ are the linewidths of the free and bound
species; $\Delta\nu_F$ is 2 Hz and $\Delta\nu_B$ corresponds to the two possible $\tau$ values
for the observed $T_{1,B}$ (see above). Using a $K_d$ of 5 x $10^{-1}$ M the
linewidths predicted by equation 8 for the two possible correlation times of the bound species are 3.5 and 17 Hz. Similarly
using a $K_d$ of 3.0 x $10^{-1}$ M linewidths of 2.3 Hz and 4.4 Hz are
predicted. Thus, it would seem that the shorter $\tau$ value for the
bound peptide would more accurately predict the observed linewidths.

Effect of Ionic Strength. Ionic strength has been shown to
be an important factor in the interaction of opiates with PS ([11]),
high ionic strength decreasing the interaction. Under conditions
of maximal interaction ("pH" 4.0), addition of 1M sodium chloride
to an enkephalin-PS mixture (1.0 mg)-(20 mg) caused the spectrum

to revert to that of free enkephalin. The high salt concentration rendered formation of vesicles impossible and spectra were obtained with the lipid simply dispersed in the mixture by sonication. A lower concentration of sodium chloride ($10^{-3}$ M) had no observable effect on the enkephalin-PS interaction.

Morphine Antagonism. In order to verify that morphine competes with enkephalin for binding to PS we added morphine sulfate pentahydrate (10 mg) to a mixture of [2-[2-$^{13}$C]glycine]enkephalin (1.2 mg) and PS (20 mg) in $D_2O$ at "pH" 6.3. Under this condition, the linewidth and intensity of the glycyl-2 resonance reverted to those of free enkephalin. Under similar conditions, but with PS (30 mg) and morphine sulfate pentahydrate (5.4 mg), the linewidth and intensity of the glycyl-2 resonance was again essentially that of free enkephalin.

Binding to Other Lipids. No interaction of [2-[2-$^{13}$C]glycine] methionine enkephalin with egg lecithin was observed at pH 3.5 and 7.0 as evidenced by no significant difference between the observed linewidths and $T_1$'s and those of free enkephalin.

Although cerebroside sulfate has been reported to interact strongly with opiates (14-16), enkephalin has so far shown no significant binding in the present type of experiment to an egg lecithin (27 mg) - sulfatide (25 mg) mixture at pH 6.3 and 7.1. Further studies are currently underway with this system.

Conclusions

The present study has shown that enkephalin binds to PS in a pH-dependent fashion. The binding most likely involves the $NH_3^+$ group of the tyrosine residue of enkephalin and the $CO_2^-$ group of PS. We have measured $K_d$ for the interaction and find it to be of the order of $5 \times 10^{-1}$ M ("pH" 6.3) which is much weaker than the interaction of PS with morphine derivatives. Upon binding of enkephalin to PS the correlation time for internal motion in the backbone of the peptide increases by at least one order of magnitude (from $7.0 \times 10^{-11}$ sec rad$^{-1}$). The $T_1$ of the bound peptide has been estimated to be 35 msec (monitored for both the glycyl-2 and glycyl-3 residues). The binding exhibits the expected morphine competition. At present no evidence of binding of enkephalin to egg lecithin or cerebroside sulfate has been obtained.

We believe this study to be preliminary to the full mapping of the enkephalin interaction with PS and to the investigation of enkephalin in high affinity binding. Future studies should involve [4-[2-$^{13}$C]phenylalanine]methionine enkephalin and [2-[2-$^{13}$C]glycine]methionine enkephalinamide, which are in preparation. These studies will allow more accurate determination of the rates of overall and internal motion in enkephalin and help elucidate further the nature of enkephalin-lipid interactions.

## References

1.  Roques, B.P., Garbay-Jaureguiberry, C., Oberlin, R., Anteunis, M. and Lala, A.K., Nature, (1976) *262*, 778.
2.  Combrisson, S., Roques, B.P. and Oberlin, R., Tetrahedron Lett., (1976), 3455.
3.  Bleich, H.E., Day, A.R., Freer, R.J. and Glasel, J.A., Biochem. Biophys. Res. Commun., (1977) *74*, 592.
4.  Fournié-Zaluski, M., Prange, T., Pascard, C. and Roques, B.P., ibid, (1977), *79*, 1199.
5.  Gorin, F.A. and Marshall, G.R., Proc. Natl. Acad. Sci., USA , (1977) *74*, 5179-5183.
6.  Smith, G.D. and Griffin, J.F., Science, (1978), *199*, 1214.
7.  Baladis, Y.Y., Nikiforavich, G.V., Grinsteine, I.V., Vegner, R.E., and Chipens, G.I., FEBS Lett., (1978), *86*, 239.
8.  Loew, G.H. and Burt, S.K., Proc. Natl. Acad. Sci., USA, (1978), *75*, 7.
9.  Humblet, C., DeCoen, J.L. and Koch, M.H.J., Arch. Int. Physiol. Biochim., (1977), *85*, 415.

10. Tancrède, P., Deslauriers, R., McGregor, W., Ralston, E., Sarantakis, D., Somorjai, R.L. and Smith, I.C.P., Biochem., (1978), *17*, 2905.

11. Abood, L.G. and Hoss, W., Eur. J. Pharmacol., (1975), *32*, 66.

12. Abood, L.G. and Takeda, F., *ibid*, (1976), *39*, 71.

13. Abood, L.G., Salem, N.Jr., MacNeil, M., Bloom, L. and Abood, M.E., Biochem. Biophys. Acta., (1977), *468*, 51.

14. Loh, H.H., Cho, T.M., Wu, Y.C. and Way, E.L., Life Sci., (1974), *14*, 2231.

15. Cho, T.M., Cho, J.S. and Loh, H.H., *ibid*, (1976), *18*, 231.

16. Loh, H.H., Law, P.Y., Ostwald, T., Cho, T.M. and Way, E.L., Fed. Proc., (1978), *37*, 147.

17. Snyder, S.H. and Simaniov, R., J. Neurochem., (1977), *28*, 13.

18. Seimija, T. and Ohki, S., Biochem. Biophys. Acta., (1973), *298*, 546.

19. Swift, T.J. and Connick, R.E., J. Chem. Phys., (1962), *37*, 307.

20. Jarrell, H.C., Tancrède, P., Deslauriers, R., McGregor, W. and Smith, I.C.P., in preparation.

21. Birdsall, N.J.M., Hulme, E.C., Bradbury, A.F., Smyth, O.G. and Snell, C.R., Opiates and Endogenous Opioid Peptides, 1976, Elsevier/North-Holland Biomedical Press, Amsterdam, p. 19.

22. Newman, G.C. and Huang, C., Biochemistry, (1975), *14*, 3363.

23. Gent, M.P. and Prestegard, J.H., *ibid*, (1974) *13*, 4027.

24. Lyerla, Jr., J.R. and Levy, G.C., Topics in Carbon-13 NMR Spectroscopy, Levy, G.C., ed., Vol. 1, 1974, Wiley, New York.

25. Deslauriers, R. and Somorjai, R., J. Amer. Chem. Soc., (1976), *98*, 1931.

RECEIVED March 13, 1979.

# Carbon-13 Spin Relaxation Parameters of Semicrystalline Polymers

D. E. AXELSON[1] and L. MANDELKERN

Department of Chemistry and Institute of Molecular Biophysics,
Florida State University, Tallahassee, FL 32306

The overall molecular structure and morphology of a semi-crystalline polymer is generally admitted to represent a very complex situation. (1)(2)(3) Crystallinity is rarely if ever complete and, depending on molecular weight and crystallization conditions, can range from about 30 to 90% for homopolymers. (1)(4)(5) Above the level of the unit cell a lamella-like crystallite is the usual predominant feature of homopolymer crystallization, and is considered to be the primary structural entity. In addition, however, the crystallites can be organized into higher levels of morphology, or supermolecular structure such as spherulites or other geometrical forms. (6)(7)(8) There are, therefore, regions of different chain structures present in different proportions, within a semicrystalline polymer. In the most rudimentary way they can be characterized as an ordered crystalline region, a relatively diffuse interfacial region and an interzonal or amorphous region, wherein the chain units are in non-ordered conformations and connect crystallites. (1)(9) The structure and amounts of these regions determine properties. The crucial question that still needs to be resolved is the detailed structure of the non-crystalline regions, the influence of higher levels of morphology on this structure and its relation to the completely amorphous polymer at the same temperature and pressure. The thermodynamic, spectral and electromagnetic scattering properties of a large number of semicrystalline polymers have been studied for many different polymers over the last few decades. (1)(4)(5) It has been generally concluded that the degree of crystallinity is a quantitative concept. (1)(3) However, the finer details of the different structures present, and their relationship to crystallite organization, are still in need of more quantitative assessment.

Carbon-13 nuclear magnetic resonance has become an important tool with which to study the microstructure and molecular dynamics

[1]Current address: DuPont of Canada Research Center, Kingston, Ontario, Canada K7L5A5

0-8412-0505-1/79/47-103-181$08.50/0

of synthetic polymers. (10)(11)  The utilization of spin relaxation techniques to study the motions and to deduce the structural
features of bulk synthetic polymers has been demonstrated.  It is
possible to observe what are essentially high resolution $^{13}$C
spectra of bulk amorphous polymers (11)(12)(13)(14)(15) and the
non-crystalline regions of polymers (13)(16)(17) by relatively
simple techniques.  The observation of $^{13}$C spectra for this purpose can be carried out with just complete scalar proton decoupling.  The major advantages in using $^{13}$C are the identification
of resonant lines for the different individual carbon atoms in the
chain, and despite the very often high viscosity of the medium, a
lack of averaging of the relaxation parameters due to spin diffusion.  To study glassy polymers or the crystalline regions,
more complicated methods using dipolar decoupling, cross-
polarization and magic angle spinning need to be used.(18)(19)(20)
      In the present work we restrict our studies to scalar proton
decoupled spectra and the determination of the spin relaxation
parameters under these experimental conditions.  We are thus
limiting ourselves at present to probing motions, and relating
them to structure, within the mobile non-ordered regions of the
semicrystalline polymers at temperatures well above the glass
temperature.  In fact only the results at relatively high temperatures will be discussed here to avoid complications caused by
transitions observed at lower temperatures. (21)  Low temperature
studies of linear and branched polyethylene as well as their
copolymers will be reported subsequently. (22)

Experimental

      Proton  decoupled natural abundance $^{13}$C Fourier transform
NMR spectra were obtained on a                      Bruker HX270 at
67.9 MHz.            Whenever possible the samples were studied
as powders but molded films were also used.  It was found that no
lock material was necessary for field frequency stabilization.
Drift tests showed that field fluctuations were orders of magnitude  smaller (< 0.5 Hz) than the typical  linewidths investigated, which were the order of several hundred Hz.  Finely
powdered poly(vinyl chloride) was used to suspend chunks or pel-
lets of polyethylene for cases of sample limitation.  The linewidths were found to be independent of the filler used as long as
temperature variations due to decoupling were minimal.  The linewidth measurements were obtained under identical experimental
conditions as possible with particular attention being paid to
noise modulation, bandwidth and decoupling field strength.  When
comparisons were important the samples were investigated consecu-
tively, if possible.  Otherwise, polymer samples of known linewidth were used as standards.  With this procedure, variations
were kept to a minimum and meaningful comparisons could be made.
Two-level decoupling was used to minimize sample heating. (23)
Quadrature detection was employed with the appropriate spectral

widths and data points to produce a well-defined resonance in the
frequency domain. Linewidths were measured directly from the
plotted spectra, with allowance being made for the artificial line
broadening caused by applying an exponentially decaying function
to the free induction decay. In some instances a computer generated least squares fitting routine was used with essentially
identical results.

Spin-lattice relaxation times were measured by the fast
inversion-recovery method (24) with subsequent data analysis by a
non-linear three parameter least squares fitting routine. (25)
Nuclear Overhauser enhancement factors were measured using a
gated decoupling technique with the period between the end of the
data acquisition and the next 90° pulse equal to about four times
the $T_1$ value. Most of the data used a delay of about ten times
the $T_1$ value. (26)

High temperature measurements were carried out using a Bruker
B-ST 100/700 variable temperature unit.

Selective saturation experiments were performed with the
HX270 employing the homonuclear decoupling mode available with the
instrument but substituting the usual $^1H$ - $^1H$ situation by
$^{13}C$ - $^{13}C$ double resonance. A Schomandl type ND100M power
amplifier provided the second frequency. A spurious peak was
generally observed at the irradiating frequency which could be
attenuated by a slight change in delay times used. However,
these have been removed from the spectra reproduced here for
purely cosmetic reasons.

The properties of all but the two lowest molecular weight
linear polyethylene samples have been previously described. (17)
These include molecular weight, degree of crystallinity and
morphological form. (8) (17)  The fraction $M_n$ = 8.6 x $10^3$ was
obtained by column fractionation and characterized in the usual
manner. (27)(28)  The sample labeled 1 x $10^3$ is also known as
Polywax 1000. It is manufactured by the Petrolite Corporation.
Its actual molecular weight, as obtained from gel permeation
chromatography, is $M_w$ = 1263; $M_n$ = 1136. (29)  Its melting temperature, under the conditions of the NMR measurements, was 108.8°C
as determined by differential calorimetry. The low density
(branched) polyethylenes studied here were commercial varieties
whose molecular weights, distribution and side group concentrations have been reported. (30)  The ethylene-butene-1 copolymers
were a gift from the Exxon Chemical Corporation.

The four polyethylene oxide samples were obtained from the
Union Carbide Corporation, and were used in the powdered form in
which they were received. Their molecular weights were obtained
in the conventional manner and are respectively $M_n$ = 3.2 x $10^4$;
$M_n$ = 2.7 x $10^4$; $M_w$ = 6.0 x $10^4$; $M_w$ = 6.67 x $10^5$; $M_n$ = 6.0 x $10^5$;
$M_w$ = 3.3 x $10^6$; $M_n$ = 1.2 x $10^6$  for the samples studied. The
enthalpy of fusion of the samples, in the original powder form,
ranged from 37.8 to 39.6 cal/g, indicating that there was not very
much difference in the level of crystallinity among the samples

in this form on this basis of calculation. If 45 cal/g is taken
as the heat of fusion for the completely crystalline sample (4),
then the degree of crystallinity is about 0.85 for these samples.

The polytrimethylene oxide sample was a gift from
Professor J. E. Mark. Its viscosity average molecular weight was
100,000.

Care was taken to check periodically, by differential scan-
ning calorimetry and small angle light scattering, whether any
changes occurred in the level of crystallinity or morphology. If
such changes occurred, then the samples were no longer used.

Results

Linewidths. The major interests in the experimental linewidths of semicrystalline polymers are how they depend on the
morphology and level of crystallinity, their variation with
temperature, and whether they are a continuous function through
the melting temperature. Experiments have been designed to
elucidate these factors and to supplement and make clearer those
previously reported. (16)(17) The temperature dependence of the
linewidth on molecular weight and level of crystallinity, of
linear polyethylene, is illustrated in Figure 1. The very high
molecular weight sample, M = 2 x $10^6$, has a degree of crystallinity, $1-\lambda$, of about 0.51, is non-spherulitic, and does not
possess any well-defined supermolecular structure. (17) For this
sample the linewidth is a smoothly monotonicly decreasing function
of the temperature. There is no indication of any discontinuity
at elevated temperatures or upon melting. An asymptotic value
for the linewidth of about 200 Hz is reached at 128°C. The intermediate sample of low molecular weight, M = 8.6 x $10^3$, with $1-\lambda$
of about 0.90 should have a rod-like morphology. (8)(31) In this
case, a resolvable spectrum cannot be obtained below 90°C. Above
this temperature the linewidth rapidly approaches the limiting
value characteristic of the much higher molecular weight sample,
M = 2 x $10^6$. Thus in the melt, or completely amorphous state,
the linewidths for these two polymers are essentially identical
despite the almost three orders of magnitude difference in molecular weight. The very lowest molecular weight sample studied,
designated as 1 x $10^3$, with a very similar level of crystallinity
and morphology as the 8.6 x $10^3$ sample does not yield resolvable
spectra until a temperature of 75-80°C is reached. Above the
melting temperature for this polymer, a constant much lower
limiting value of about 50 Hz is found for the linewidth. Since
in this case spectra cannot be obtained below 75°C, a clear
decision cannot be made as to whether a discontinuity in the
linewidths occurs upon melting. The data are suggestive, however,
that a discontinuity does in fact exist. In the molten state
there is a very large difference in linewidth for M = 1 x $10^3$ as
compared to the other samples. However, the asymptotic value has

already been obtained for only a very slightly higher molecular
weight, M = 8.6 x $10^3$.

The influence of morphology on the linewidth-temperature
relations is illustrated in Fig. 2 for two high molecular weight
samples which have the same degree of crystallinity at room
temperature. (17) The major influence of the morphology on the
observed linewidths, that have been previously reported (17),
manifests itself below 90°C. As the temperature is decreased the
linewidth differences become continuously greater between the two
samples. In the vicinity of 30-40°C the differences correspond
to the values previously reported. (17) Above 90°C the linewidths of the two samples are indistinguishable from one another,
and both have the same value in the melt. The continuous character of both curves indicates that either the fusion process with
regard to morphology is different in the two cases; or if the
morphological forms are maintained, the structures in the non-
crystalline regions possess a different temperature dependence.
More detailed studies which relate any changes in morphology with
the course of fusion for these kinds of samples are necessary to
resolve these points. For present purposes, the major, definitive
conclusions are that the temperature dependence of the linewidth
is continuous through the melting temperature for these two
extremes in morphology. Moreover, depending on the morphology,
significant differences in the magnitude of the linewidth are
observed below 90°C.

Fig. 3 represents a similar phenomenon for two samples which
possess spherulitic morphologies but have different levels of
crystallinity. At the intermediate temperatures the increased
degree of crystallinity yields only a slightly greater linewidth
for the same morphology. This small difference remains constant
until the melting is reached, at which point the two curves
merge. At the lower temperatures, the same high value characteristic of spherulites and independent of the level of crystallinity is observed as was previously reported. (17)

The plot in Fig. 4 represents in the same diagram the
influence of both the degree of crystallinity and morphology,
from examples taken from the previous two figures. As would be
anticipated from the previous results, the differences in the
linewidths below the melting range are greatly enhanced when the
data are examined in this manner. However, as is equally clear,
the results would be difficult to sort out for interpretive
purposes if a clear distinction was not made between the levels
of crystallinity and morphology, or lack thereof, as distinctly
different, independent quantities.

The linewidth-temperature relation of the polyethylene
oxide samples are given in Fig. 5. Despite the large differences
in molecular weight, these samples have about the same linewidth,
300-350 Hz, in the crystalline state at 25°C. They all also
possess a spherulitic type of morphology. The influence on the
linewidth of the different types of supermolecular structures

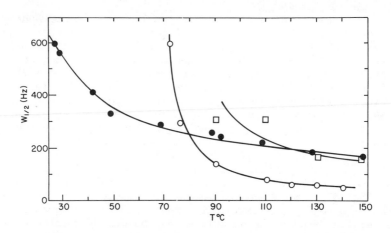

*Figure 1.   Plot of linewidth, $W_{1/2}$, against temperature at 67.9 MHz for linear PE's of indicated molecular weights: ($\bullet$), $2 \times 10^6$ mol wt; ($\square$), $8.6 \times 10^3$ mol wt; and ($\bigcirc$), $1 \times 10^3$ mol wt.*

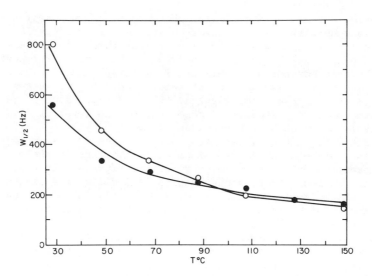

*Figure 2.   Plot of linewidth, $W_{1/2}$, against temperature at 67.9 MHz for two linear PE samples of same degree of crystallinity (0.51) but differing morphologies: spherulitic, ($\bigcirc$); no morphology ($\bullet$).*

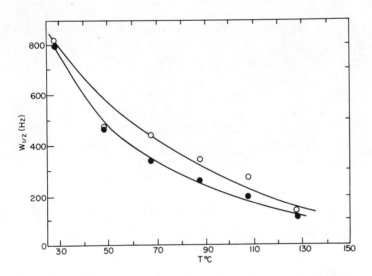

*Figure 3.   Plot of linewidth, W$_{1/2}$, against temperature at 67.9 MHz for two linear PE samples having a spherulitic morphology but differing levels of crystallinity: degree of crystallinity 0.78, (○); degree of crystallinity 0.50, (●).*

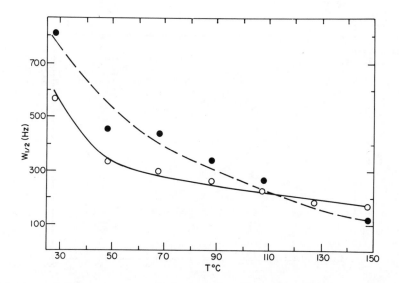

*Figure 4.   Plot of linewidth, $W_{1/2}$, against temperature at 67.9 MHz for two linear PE samples of differing morphologies and levels of crystallinity.   No morphology, degree of crystallinity 0.50, (○); spherulitic morphology, degree of crystallinity 0.78, (●).*

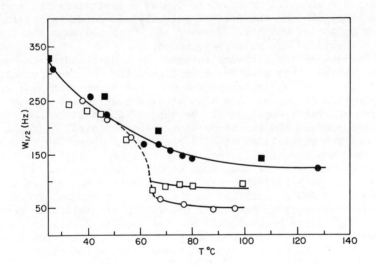

*Figure 5.   Plot of linewidth, $W_{1/2}$, against temperature at 67.9 MHz for PEO samples of indicated molecular weights: (■), $3.3 \times 10^6$ mol wt; (●), $6 \times 10^5$ mol wt; (□), $6 \times 10^4$ mol wt; (○), $3 \times 10^4$ mol wt.*

that can be developed in polyethylene oxide has not as yet been investigated. For the two highest molecular weight samples studied here, the linewidths are a continuously decreasing function of temperature. A limiting value of approximately 135 Hz is attained at the elevated temperatures in the melt. There is no evidence of any discontinuity in this quantity in the melting region of 60-70°C. However, although all four samples have essentially the same linewidth in the crystalline state, there is a definite discontinuity upon melting for the two lowest molecular weight samples. The change in linewidth is greatest for the lowest molecular weight sample and is independent of temperature above its melting temperature. The limiting linewidths in the molten state are 80-90 Hz for $M = 6 \times 10^4$ and are reduced to 47 Hz for $M = 3 \times 10^4$. In contrast to linear polyethylene the asymptotic linewidth for the melt of polyethylene oxide is attained at higher molecular weights. However, for both polymers the asymptotic linewidths are relatively broad.

Because of the ultimate interest in understanding the segmental motions that contribute to linewidths and the apparently broad lines that are observed for both linear polyethylene and polyethylene oxide, we have also studied a series of random ethylene-butene-1 copolymers. Because of the high co-unit content of these samples and the temperature range studied, if these polymers are crystalline at all, the level is very low. Above 50°C they are essentially completely amorphous. (32) The line-width measurements for four such copolymers are given in Fig. 6. Above 50°C the linewidths only change slightly and are in the range of 180-200 Hz, similar to the melt of the homopolymer. As the temperature is lowered well below 35°C crystallinity begins to develop. This problem will be discussed in detail in a subsequent paper. (22) For present purposes, for these copolymers, we are mainly concerned with the linewidths in the amorphous state.

Spin-Lattice Relaxation Parameters, $T_1$. In Fig. 7 the $NT_1$ values for polyethylene oxide samples are plotted as a function of temperature. Also plotted are the average values for the two different carbons of polytrimethylene oxide. Except for the very lowest molecular weight polyethylene oxide sample (which corresponds to a degree of polymerization of only 4.5), the spin-lattice relaxation time is the same for all the samples at a given temperature and is also a continuous function of the temperature. The respective melting temperatures of each of the polymers is indicated by the vertical arrows in the figure. It thus becomes abundantly clear from this extensive set of data that the $T_1$'s are continuous functions through the melting temperature. The segmental motions of the non-crystalline regions which are represented by this parameter are not dependent on the state of the polymer. The lowest molecular weight polyethylene

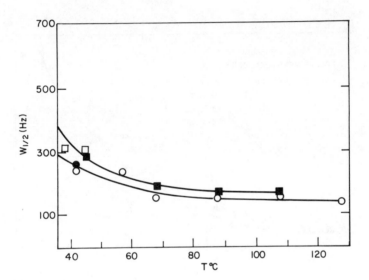

*Figure 6.   Plot of linewidth, W₁/₂, against temperature at 67.9 MHz for ethylene-butene-1 random copolymers. Co-unit contents: (■), 10; (□), 7; (●), 21; (○), 26.*

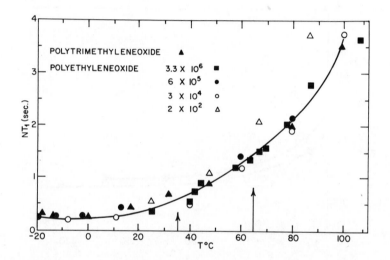

*Figure 7.  Plot of spin-lattice relaxation time against temperature, at 67.9 MHz,
for polytrimethylene oxide and PEO.*

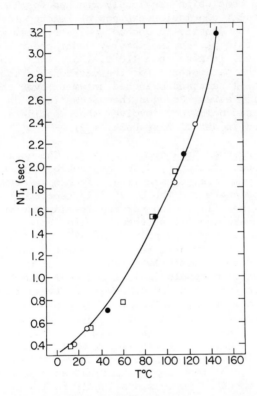

*Figure 8.   Plot of spin-lattice relaxation time against temperature, at 67.9 MHz, for linear and branched PE.   Linear PE from Ref. 17, (●); linear PE this work, (○); branched PE this work, (□).*

oxide sample gives significantly higher $NT_1$ values at the elevated temperatures.

A similar continuity in the $^{13}C$ $T_1$'s through the melting temperature was previously reported for linear polyethylene. (17) We have now investigated the temperature dependence of this quantity, for this polymer, in more detail and have also studied a low density (branched) polyethylene. The results for the polyethylenes are summarized in Fig. 8. The new data reported here substantiate the conclusion previously reached for linear polyethylene. A similar conclusion can now be reached for the backbone carbons of low density (branched) polyethylene. The melting temperature for this particular sample, under the crystallization conditions studied, is less than 110°C. (33) Thus, the $^{13}C$ spin-lattice relaxation parameters for the backbone carbons are the same for both the linear and branched polymers over the temperature range studied here. Changes that occur in $T_1$ as the temperature is reduced below 0°C involve other considerations and will be discussed in detail elsewhere. (22)

Nuclear Overhauser Enhancement Factor. The nuclear Overhauser enhancement factor, NOEF, has not been as extensively studied for the pure semi-crystalline polymers as have the other $^{13}C$ relaxation parameters. The previously reported results for linear polyethylenes (17), with varying levels of crystallinity and different supermolecular structures, are included in the summary of Table I. For this polymer at 45°C and 67.5 MHz, within the limits of the experimental results presently available, the NOEF appears to depend primarily on the level of crystallinity and not on any other morphological, structural or molecular factors. The samples with the lowest level of crystallinity that could be attained with this polymer, 0.50, possess full NOEF's which are also found in the pure melt. (34) For the higher crystallinity samples the NOEF is reduced, leading to the suggestion that the degree of crystallinity could be an important factor in determining this quantity. We have explored the temperature dependence of the sample which had an NOEF of 1.5 at 45°C. Although the NOEF increased slightly to 1.7 at 100°C, it has not yet achieved its maximum value at this point. Since the full NOEF must be attained in the pure melt, it is not as yet clear whether or not the change will be continuous.

The NOEF's for two of the polyethylene oxide samples, $M = 3.2 \times 10^4$ and $6.67 \times 10^5$, were determined, as a function of temperature, from room temperature through the melting point. In both cases full NOEF's were observed at room temperature and in the melt. Schaefer and Natusch have reported essentially a full NOEF, $1.7 \pm 0.1$, for a 1000 molecular weight polyethylene oxide which is a liquid at room temperature. (35)

It is clear that more extensive studies are needed before any detailed interpretation can be given to the NOEF's, or trends discerned, relative to the amorphous structure of the

Table I.

Carbon-13 Spin Relaxation Parameters of Linear Polyethylene (17) at 45°C and 67.9 MHz

| $M_\eta$ | $(1-\lambda)_d$ | Morph. | $T_1$ (msec)[a] | Linewidth $W_{\frac{1}{2}}$ (Hz)[b] | NOEF |
|---|---|---|---|---|---|
| $8.1 \times 10^4$ | 0.57 | Spher. | 343 | 622 | – |
| $2.5 \times 10^5$ | .51 | Spher. | 355 | 625 | 2.0[c] |
| *$1.7 \times 10^5$ | .81 | Spher. | 348 | 695 | 1.5[d] |
| *$1.7 \times 10^5$ | .68 | Spher. | 356 | 700 | – |
| *$2.0 \times 10^6$ | .51 | None | 369 | 501 | 2.0[c] |
| *$2.0 \times 10^6$ | .72 | None | 358 | 496 | – |
| $6.1 \times 10^6$ | .54 | None | – | 500[e] | – |
| $6.1 \times 10^6$ | .70 | None | – | 503 | – |
| $2.75 \times 10^4$ | .94 | Rod | 352 | 945 | 1.0[d] |

*Unfractionated sample
[a] estimated accuracy ±10%
[b] estimated accuracy ±5–10%
[c] estimated accuracy ±0.1
[d] estimated accuracy ±0.2
[e] estimated

semicrystalline polymers. However, there are certain inherent
restraints which make difficult the determination and collection
of as much data as would be desired. There are no problems with
systems which can be described by a one-correlation time model,
such as linear polyethylene of low levels of crystallinity (17)
and the polyethylene oxide samples just described. However, when
a distribution of correlation times are necessary to describe the
results, complications can exist. For example, even in the completely amorphous state cis polyisoprene at 40° and 0°C possesses
low NOEF's which are very close to the theoretically allowed
minimum. These results are a consequence of the distribution of
relaxation times that are required to describe the system.
(11)(13)(14)(16) Consequently, upon crystallization very little
change can be expected, as is observed experimentally. (16) Thus
the number of systems that can be effectively studied, to assess
the influence of crystallinity on the NOEF's, are limited. These
will have to be intensively studied in more detail before any
general conclusions can be deduced.

Discussion

    Spin-Lattice Relaxation Parameters. As is summarized in
Table I, it was previously shown for linear polyethylene that at
fixed temperature, the $T_1$'s are constant for a wide range in the
level of crystallinity and molecular weight and for different
morphologies or lack thereof. (17) The levels of crystallinity
varied from 0.50 to 0.94 in this study. Thus the segmental
motions, reflected in $T_1$, are dependent only on the properties of
the amorphous state. This conclusion is further substantiated by
the temperature dependence of $T_1$, which is illustrated in Fig. 8.
These data clearly demonstrate the continuity of $T_1$ through the
melting temperature for both linear and branched polyethylene.
Consequently as far as the $T_1$ measurements are concerned, the
segmental motions from the non-crystalline regions that contribute
to this quantity are identical to those of the pure melt. The
amount of crystallinity, the organization of the crystallites and
the temperature, therefore, have no influence on the high frequency segmental motions in the non-crystalline regions.
    These conclusions are further generalized by the more exten-
sive data presented in Fig. 7 for polyethylene oxide and polytrimethylene oxide. The continuous nature of the $T_1$ function for
both these polymers over a large temperature range is quite
definite and is emphasized by the detailed data in the vicinity of
the respective melting temperatures. This is true even for the
polyethylene oxide samples where discontinuities in the linewidth
are clearly indicated in Fig. 7. Obviously, the type of segmental
motions which contribute to the two different relaxation parameters are influenced quite differently by the presence of
crystallinity.
    The $T_1$ values for the very low molecular weight polyethylene

oxide sample are essentially the same as the high molecular weight ones at low temperatures.  However, above 50°C the $NT_1$ increases. Since the degree of polymerization for this sample is very low, being only a few repeating units, overall molecular reorientation can take place at these elevated temperatures.  Thus individual molecular contributions are being made.  A similar behavior for low molecular weight polyethylene oxide has been reported (36) as have proton $T_1$ measurements for low molecular weight polyethylene. (37)

The major results described could be partially anticipated from those previously reported for linear polyethylene (17) as well as those for cis polyisoprene. (16)  For the latter polymer, by taking advantage of its crystallization kinetic characteristics, it was possible to compare the $^{13}C$ relaxation parameters of the completely amorphous and partially crystalline polymer (31% crystallinity) at the same temperature, 0°C.  This is a unique situation and allows for some unequivocal comparisons.  It was definitively observed that for all the carbons of cis polyisoprene the $T_1$'s did not change with crystallization.

Based on the results obtained to date, which have been summarized above for several different semicrystalline polymers—linear and low density (branched) polyethylene, polytrimethylene oxide, polyethylene oxide and cis polyisoprene—it is concluded that the relatively fast segmental motions, as manifested in $T_1$, are independent of all aspects of the crystallinity and are the same as the completely amorphous polymer at the same temperature. Furthermore, it has previously been shown that for polyethylene, the motions in the non-crystalline regions are essentially the same as those in the melts of low molecular weight n-alkanes. (17)

Proton NMR relaxation parameters have also been determined for polyethylene (38) and polyethylene oxide (39) in the melting region.  The apparent contradiction between the proton spin-lattice relaxation parameter for a high molecular weight linear polyethylene sample at its melting point, with the $^{13}C$ relaxation measurements, has previously been pointed out. (17)  This discrepancy is still maintained with the more detailed results reported here for both types of polyethylene.  For the proton relaxation a small, but distinct, discontinuity is reported at the melting temperature. (38)

On the other hand, Connor and Hartland (39) have reported results of a proton NMR study for a series of polyethylene oxide samples by rotating frame proton relaxation time $T_{1\rho}$ measurements. $T_1$ was also determined.  For their lowest molecular weight sample, M = 550, the $T_{1\rho}$ values display a fairly sharp discontinuity at about 12°C.  This temperature is close to the independently determined melting range of 15-25°C.  However, there is no evidence for a discontinuity in the $T_1$ measurements for this sample. Above the melting temperature the values found for $T_1$ and $T_{1\rho}$ were similar.  $T_{1\rho}$ exhibited a very pronounced discontinuity at about 63°C for the sample M = 6000.  This discontinuity occurs within the reported melting range of 60-63°C. Although there is a minimum of $T_1$ in this temperature region, there is no indication of any discontinuity. In contrast to these two lower molecular weight samples, for a high molecular weight sample that was also studied, M = 2.8 x 10$^6$, no sharp discontinuity was observed in $T_{1\rho}$ in the vicinity of the melting temperature and $T_1$ was continuous. In summary, for the two lowest molecular weight samples, $T_{1\rho}$ displayed a discontinuity in the melting region while $T_1$ was continuous. For the very high molecular weight sample both of these relaxation parameters were continuous with temperature. These results are reminiscent of and very similar to those reported here for the $^{13}C$ relaxation parameters of polyethylene oxide as illustrated in Figs. 5 and 7. We have found that the $T_1$'s are continuous with temperature over a very wide molecular weight range. However, as is shown in Fig. 5, there are discontinuities in linewidths at the melting temperatures for the low molecular weight samples, but the linewidths are continuous for the higher molecular weight samples. The $T_{1\rho}$ results of Connor and Hartland (39) display virtually the same behavior. These similarities do not appear to be coincidental since the linewidth is directly related to the spin-spin relaxation parameter, $T_2$, which is sensitive to low frequency motions. The $T_{1\rho}$ values are also more sensitive to the lower frequency motions and would thus be expected to behave similarly to $T_2$, or linewidths, in the $^{13}C$ relaxation studies.

     Linewidths. Ideally the spin-spin relaxation time, $T_2$, can be obtained from the linewidth $W_{\frac{1}{2}}$ by the relation $T_2 = 1/(\pi W_{\frac{1}{2}})$. Although in the present work we are mainly interested in the influence of the different aspects of crystallinity on $T_2$, it is informative to examine the linewidth in the completely amorphous state above the melting temperature. This state conveniently serves as a reference point for the subsequent discussion. As we shall deduce in the subsequent discussion, it is also very important in establishing an understanding of the conditions for a discontinuity to be observed in the vicinity of the melting temperature. We have previously reported (15) that for non-crystalline polyisobutylene samples, at 45°C and 67.9 MHz, the linewidths for all the carbons are independent of chain length for molecular weights greater than about 4-5 x 10$^4$. Below this molecular weight the linewidths are substantially reduced with decreasing chain length. The limiting, or leveling-off, value for the methylene carbons is about 200 Hz for this polymer. For linear polyethylene, we observed a very similar effect, as was illustrated in Fig. 1. In the pure melt above the melting temperature at 140°C, and at the same frequency, a limiting value of about 200 Hz for the linewidth is observed. The limiting value is attained at a much lower molecular weight for this polymer under these conditions. For the very low molecular weight sample, M = 1 x 10$^3$, the linewidth is substantially reduced to about 50 Hz. The completely amorphous ethylene-butene-1 copolymers, whose molecular weights are in the asymptotic region, have linewidths of the order of 180-200 Hz above 50°C. These are very similar to the completely amorphous homopolymer. The results for polyethylene oxide follow the same patterns. An asymptotic linewidth of 135 Hz is reached in the molten state between $6 \times 10^4$ and $6 \times 10^5$. A monotonic decrease in linewidth is observed below this molecular weight. For high molecular weight cis polyisoprene, in the completely amorphous state at 40°C and at 67.9 MHz, the linewidths for all the carbons are only about 40 Hz. (16) Although additional type polymers should be studied before firm generalizations can be made, the data in hand indicate certain salient features relative to the amorphous states. There is a low critical molecular weight, whose exact value varies with polymer type, above which the linewidth and thus $T_2$ is independent of chain length. This behavior has now been observed with the three different polymers studied in detail and could be expected to be universal. Except for cis-polyisoprene, the linewidths for the other polymers are relatively broad compared to other type molecular systems. The rather narrow lines observed for the methylene carbons of cis polyisoprene, about 40 Hz at 67.9 MHz and 20 Hz at 22.9 MHz (16), are also found in the melt of other diene polymers. (22) They would appear from the data obtained so far to be atypical of polymers and a consequence of the double bond in the chain.

In contrast to the spin-lattice relaxation parameters, which remain invariant, a substantial broadening of the resonant lines occurs upon crystallization. The effect is relatively modest for cis polyisoprene at 0°C and 67.9 MHz, where comparison can be made at the same temperature. Here there is about a 50% increase in the linewidths upon the development of 30% crystallinity. Schaefer (13) reports approximately 3- to 5-fold broader lines (but they are still relatively narrow) for the crystalline trans polyisoprene relative to the completely amorphous cis polyisoprene at 40°C and 22.6 MHz. It is interesting to note in this connection that for carbon black filled cis polyisoprene the linewidths are greater by factors of 5-10 relative to the unfilled polymer.

In the present work the limiting value of the linewidths for polyethylene oxide increases from 135 Hz in the melt above 70°C, to the range 300-350 Hz in the crystalline state at room temperature. As is indicated in Table I, the resonant linewidths for linear polyethylene increase substantially upon crystallization and attain values in the range 500-900 Hz at 45°C and 67.9 MHz. As has been emphasized previously (17), the level of crystallinity is not the major determinant of the linewidth in the semicrystalline state. Rather the supermolecular structure or morphology is a major factor in governing the magnitude of the linewidth. Structural factors and crystallization conditions under which low density (branched) polyethylene forms

either spherulites or no well-defined morphology have recently been
established. (33) Samples from each of these structural cate-
gories yield linewidths which are virtually identical with the
corresponding values listed in Table I for the linear polymer. A
morphological map, similar to those developed for linear and
branched polyethylene (8)(33), has not as yet been completed for
polyethylene oxide. Thus a further generalization of these
important findings to this polymer cannot be made as yet.

The results that have been obtained indicate that the major
influence of the crystalline regions on segmental motions, and
hence to the structure of the non-crystalline regions, is in the
linewidth and $T_2$. The different morphologies are reflected in
different values of $T_2$. The segmental motions in long chain
molecules which exert major influence on the spin-lattice relaxation times and the nuclear Overhauser enhancements are not in
general the same motions which determine the resonant linewidth.
$T_1$ is in general greater than $T_2$. This difference can in part
be a consequence of the slower modes of polymer motion, which are
characterized by correlation times sufficiently long that they
do not contribute significantly to $T_1$ but do to $T_2$. It is therefore important, in terms of describing the fine structure of the
non-crystalline regions, to understand the type motions which
contribute to $T_2$ and to develop a rationale for the relatively
broad lines that are observed for most crystalline polymers.

There are many possible factors that can contribute to the
linewidths. It is important, therefore, that the pertinent ones
be discerned and understood if the molecular interpretations are
to be eventually deduced. The task is a formidable one and the
complete solution of the problem is not as yet in hand. Many
of the possible contributions to the linewidth, and reasons for
the apparently excessive broadening, have been previously discussed. (11)(13)(17) Several mechanisms have emerged as being
most likely contributors. One must, however, be willing to
accept the concept that several different mechanisms can be
simultaneously involved and contributing to the line broadening.
Therefore it is necessary that a diverse set of experiments be
designed and carried out, to substantiate or dismiss the different
possibilities. A unique process is not necessarily required and
should not be established as an objective. A large number of
different type experiments are necessary to sort out the different possibilities. In consideration of the amount of work
involved, the problem has not as yet been completely resolved.

Line broadening due to inhomogeneity in the static magnetic
field, Ho, as well as in the rf pulse $H_1$, can contribute to the
observed resonance. However, studies of standard samples, of
known natural linewidths, enable the contributions from this
source to be determined. In the present case these causes contribute only a few percent, i.e., a few Hz, to the total linewidth
and are thus inconsequential to the present problem. Before
discussing the different motional contributions to the linewidth,

the possibility that non-motional or static phenomena can also make a substantial contribution needs to be given serious consideration. (40)(41) Differences in bulk magnetic susceptibility within the same volume element can result in differences in nuclear screening among nuclei in different regions of the sample, resulting in a broadening of the resonance lines. (40)(41)(42) Although such broadening can occur from the irregular macroscopic sample configuration, it most likely arises from microscopic structural differences within the sample. Broadening from this cause alone will vary linearly with the applied field. (41) Thus, because of the high field used in the present work, if this process were operative, it could be quite severe, and would have escaped notice in most previous studies which have been conducted at much lower field strengths. Consequently, we have carried out a detailed study of the frequency dependence of the linewidths for the polymers studied here. Particular attention has been given to the influence of crystallinity, morphologic form and temperature. An extensive set of data has now been collected and analyzed. We shall limit ourselves here to a brief summary of the major findings as they pertain to the major themes of the present work. A more detailed report of these findings will be presented elsewhere. (43)

A substantial effect of the field on the resonant linewidths was found for the crystalline polyethylenes and polyethylene oxides. The magnitude of the changes with frequency is in qualitative accord with theoretical expectation. If other molecular and constitutional factors are held constant, then the influence of the morphology on the linewidth, which was previously observed at 67.9 MHz, is still maintained at the lower frequencies. Thus, the low frequency segmental motions of the noncrystalline regions are definitely influenced by the morphology; the previous conclusion was not a consequence of the high fields that were used. This fact thus has important molecular implications with respect to the structure. For these samples the extrapolation of the linewidth to zero frequency does not pass through the origin. Rather large residual values, about 100 Hz for the crystalline polyethylenes and about 25 Hz for crystalline polyethylene oxides, are found. This result is consistent with a residual dipolar coupling contribution to the resonant linewidth. Preliminary magic angle spinning experiments that we have performed with crystalline polyethylene oxide at low spinning frequencies substantiate that the field-dependent broadening has a major static contribution from microscopic inhomogeneities. Thus, there are at least several contributions to the resonant linewidth and its broadening.

The homopolymers of low levels of crystallinity, as well as the ethylene-butene-1 copolymers, which are either completely amorphous, or slightly crystalline at the temperatures of measurement, also display frequency-dependent linewidths. Although these effects are not nearly as severe as in the more crystalline samples, they are not easily understood. Residual dipolar couplings could readily account for a portion of the linewidths in these cases.

The frequency dependence of the linewidth and the contributions from microscopic inhomogeneities strongly suggest significant inhomogeneous broadening. If these contributions are not completely averaged in an experiment, they will give rise to a distribution of chemical shifts and an inhomogeneous resonant line. (13) Bloembergen, Purcell and Pound (44) have shown that single frequency irradiation of a line whose width is dominated by magnetic field inhomogeneities results in the local saturation of the line. This is the so-called "hole burning" experiment and has been carried out successfully by Schaefer for several polymer systems. (13) In this experiment it is possible to determine the natural dipolar linewidth in the presence of macroscopically or microscopically inhomogeneous magnetic fields. (45) However, when attempting to saturate a resonance whose width is determined by strong spin-spin interactions, rather than field inhomogeneities, the entire line becomes saturated. In this situation the energy absorbed by the spins is no longer localized; instead, the temperature of the spin system as a whole is raised. This situation is illustrated in Fig. 9 for dioxane (15% acetone-$d_6$, 85% dioxane, ambient temperature, 67.9 MHz). This homogeneous resonant line was chosen for illustration and for comparison with a linear polyethylene sample. The spectra on the right demonstrate that an increase in the saturating rf field causes a decrease in the intensity of the resonance. Simultaneously, however, the location of the line, i.e., the point of maximum intensity, as well as the linewidth remain constant, because of the homogeneity.

The results of this type experiment for a linear, non-spherulitic polyethylene sample are shown in Figs. 10 and 11. In Fig. 10 the rf irradiation was applied at approximately the location of the maximum in the line intensity. The power levels were progressively increased to saturate the major portion of the resonance. An irregularly shaped resonance is observed. Fig. 11 demonstrates more clearly the inhomogeneous nature of the initial polyethylene resonance. In this instance, the position of the irradiating rf is progressively moved upfield, as is indicated by the vertical arrows, from the position of maximum line intensity in the unperturbed spectrum. Different power levels were used concurrently. The uppermost spectrum in Fig. 11 was obtained with no selective rf irradiation being applied and is the reference against which the remaining spectra should be compared. It is again apparent from these results that the resonant line is inhomogeneous since the symmetry, linewidth at half-height, and peak maxima change with the position and strength of the irradiation.

The inability to "burn" a narrow hole in the polyethylene spectrum is an indication that the "natural homogeneous linewidth"

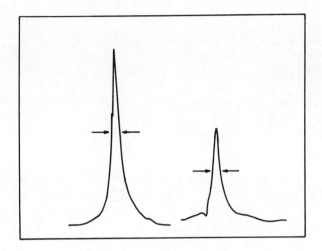

*Figure 9. Selective irradiation of dioxane. Homonuclear irradiation of a homogeneously broadened resonance at 67.9 MHz. Spectral details: PW = 40 μsec (90°C), D2 = 10 sec, 4 scans accumulated, quadrature detection; 15% acetone-d₆, 85% dioxane mixture.*

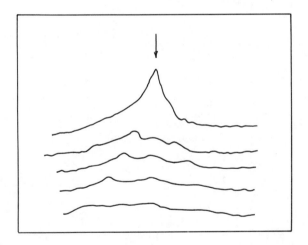

*Figure 10. Selective irradiation of linear PE (2 × 10⁶ mol wt, 1 − λ ∼ 0.5). Spectral details are: 35°C; 67.9 MHz; sweep width ± 5 KHz (quadrature detection); line broadening 9.7 Hz; pulse width 35 μsec (90°C = 48 μsec); delay = 1.0 sec, 4K data points; 1024 scans accumulated; 10-mm sample tube. Decoupling: 7W (forward), 0.4W (reflected), broad band noise modulated decoupling.*

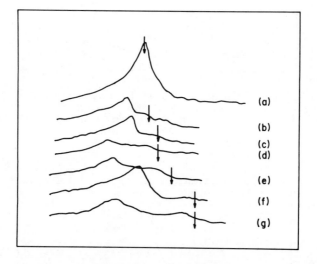

*Figure 11.  Selective irradiation of linear PE.  Spectral details: 35°C, 67.9 MHz; variable power levels.  Offsets in Hz from center frequency are: (a) 0 Hz; (b) 100 Hz; (c) 300 Hz; (d) 300 Hz; (e) 700 Hz; (f) 1300 Hz; and (g) 1300 Hz.  Same conditions as in Figure 10.*

for the particular sample studied is quite large. This conclusion is consistent with the large linewidth obtained by extrap-
olating the frequency data to zero frequency.

On the other hand, Schaefer ($\underline{13}$) has shown from selective
saturation experiments of amorphous cis polyisoprene, crystalline
trans polyisoprene, as well as carbon black filled cis polyisoprene, that the resonant lines are homogeneous. The linewidths
in these cases are thus not caused by inhomogeneous broadening
resulting from equivalent nuclei being subject to differing local
magnetic fields. The results for these systems are thus contrary
in part to what has been found here.

At this point it has been established that there are at
least two basic mechanisms which contribute to the broad lines
that are observed for the crystalline polymers. The residual
zero frequency line broadening component can be analyzed in more
detail. Specific attention can be given to factors which are a
consequence of the chain-like character of the molecules. The
local field at a given nucleus is the sum of the individual
fields contributed by the neighboring magnetic nuclei. Segmental
motions will induce a time dependence to the variables so that
the individual contributions can be described by the equation: ($\underline{46}$)

$$H_{ij}(t) = \pm \mu_{ij}/r^3_{ij}(t) \left[ 3 \cos^2\theta_{ij}(t) - 1 \right] \tag{1}$$

If $r_{ij}$ is assumed to be constant, i.e., the directly bonded
protons provide the dominant contribution, and $\theta_{ij}(t)$ is only
time dependent, the time averaged local field is given by

$$(\mu_{ij}/r^3_{ij}) \int_0^{T_2} (3 \cos^2\theta_{ij}(t)-1) \, dt. \tag{2}$$

Here $T_2$ is the order of time in which the nucleus resides in a
given spin state. If there are no restrictions on the directions
available to the internuclear vector, then the time average can
be replaced by a space average, with the result that

$$(\mu_{ij}/r^3_{ij}) \int_0^\pi (3 \cos^2\theta_{ij}-1) \sin \theta d\theta = 0. \tag{3}$$

For non-viscous liquids, where the above condition will be
expected to be fulfilled, narrow resonances are observed, when
only this source of line broadening is involved.

Other types of restrictions can be thought of being imposed
on the parameters of Eq. (1), when polymeric systems are involved.
For example, the orientation angles may be unrestricted with
respect to accessibility but the segmental motions may not be
sufficiently rapid to average $\theta_{ij}$ over the angular range 0 to $\pi$
in the time interval $T_2$. Conversely the motions may be relatively
rapid but the angular range may be restricted. It has been
calculated that excluding $\theta_{ij}$ from but a few orientations, i.e.,
excluding the magnetization vectors from solid sectors of only a
few degrees, is sufficient to produce broadening by about an order

of magnitude. (13)(46)   This latter process has been termed
incomplete motional narrowing.   It was also noted (13) that in a
selective saturation experiment a partially motionally-narrowed
line can be expected to behave as a single dipolar-broadened NMR
line.

Schaefer (13)(47) has interpreted the linewidths of cis
polyisoprene in this context and concluded that the data could be
explained by assuming that not all spatial orientations were
accessible to the chain units as a consequence of restrictions
imposed on the segmental motions.   Chain entanglements were
postulated to be the major source of these restrictions for this
amorphous polymer.   In the carbon black filled cis polyisoprene,
the filler itself was considered to be an additional source of
entanglements.   For  semicrystalline polymers,  the presence of
crystallites and their relative arrangement could play a similar
role as well as introducing inhomogeneities into the system which
can serve as another source of broadening.

The line broadening caused by partial motional narrowing can
be distinguished from that due to isotropic reorientation at a
reduced rate by appropriate magic angle spinning experiments.
(13)(47)   Random isotropic motion at reduced rates covers frequencies of the order of the inverse of the correlation time, i.e.
of the order of $10^5$-$10^7$ Hz.   Hence, sample rotation at the usually
accessible rates of $10^2$-$10^4$ Hz at the magic angle will have no
influence since the linewidths are determined by frequencies
several orders of magnitude greater.   Partial motional narrowing,
however, results in the linewidth being determined in part by
very low or zero frequency components.   These are affected by
fast magic angle spinning.   The extent of line narrowing that is
obtained depends on the distribution of the frequencies generated
by the residual dipolar interaction relative to the spinning
frequency.

The concept of chain entanglements influencing the linewidths, or $T_2$'s, can be examined more directly by studying the
influence of molecular weight.   It is well established that the
zero shear bulk viscosity of all amorphous polymers is directly
proportional to the molecular weight below a critical low molecular weight, $M_c$, and above this molecular weight increases as
the 3.5 power of M. (48)(49)(50)   $M_c$ represents approximately
twice the molecular weight between chain entanglements.   The molecular weight dependence of the viscosity results from the fric-
tional loss in the two different molecular weight regions. (48)

This molecular weight dependence manifests itself quite
distinctly in the [13]C linewidths in the completely amorphous state
and is consistent with certain proton NMR $T_2$ studies that have
been reported.   For polyisobutylene the [13]C linewidths are
invariant at 45°C for each of the carbons for molecular weights
greater than 4.5 x $10^4$ over a 60-fold change in molecular weight.
(15)   This range corresponds to a change of eight orders of

magnitude in the bulk viscosity. There is a decrease in the line-
widwth below this molecular weight. From bulk viscosity measurements, $M_c$ is found to be 1.5 x $10^4$ (51)(52), which is the same
order of magnitude that was found for the invariant linewidth.
The polyethylene linewidths, as illustrated in Fig. 1, clearly
indicate that a constant value is attained in the melt between
1-8 x $10^3$. Proton $T_2$ measurements at 150°C indicate that there is
a change in slope in this quantity as a function of molecular
weight at about $M_n$ = 6 x $10^3$. (37) The critical molecular weight
as determined from bulk viscosity measurements for this polymer
has been given as 2 x $10^3$ (53) and 3.8 x $10^3$. (52) These values
are very close to those which would be deduced from the linewidth
measurements. For polydimethyl siloxane the break in the proton
$T_2$-molecular weight curve occurs at about M = 5 x $10^4$. (37) $M_c$
is about 2.5 x $10^4$ from viscosity measurements. (52) Fig. 5
indicates that for polyethylene oxide at 90°C the linewidth
becomes constant between 6 x $10^4$ and 6 x $10^5$. The viscosity data
indicates that the critical molecular weight is about $10^4$.
(54)(55)(56) For all the cases cited above, which represent those
data for which a comparison can be presently made, there is a
direct connection between the critical molecular weight representing the influence of entanglements on the bulk viscosity and
other properties, and the NMR linewidths, or spin-spin relaxation
parameters of the amorphous polymers. Thus the entanglements
must modulate the segmental motions so that even in the amorphous
state they are a major reason for the incomplete motional narrowing, as has been postulated by Schaefer. (13) This effect
would then be further accentuated with crystallization.

From the above, the observations that in some cases a discontinuity at the melting temperature is observed in the [13]C
linewidths, while in other situations the linewidth is continuous,
can be readily explained. For polyethylene oxide the linewidths
in the crystalline state, for the samples studied to date, are all
about the same, presumably due to the similarity in morphology
and level of crystallinity. However, due to the differences in
the linewidth in the amorphous state, the lower molecular weight
samples must exhibit a discontinuity at the melting temperature,
while the higher ones will be continuous. A similar situation
will exist for polyethylene. In this case it has been seen in
Fig. 1 that the major influence of morphological form on the
linewidth at lower temperatures has disappeared for the high
molecular weight sample. Thus it is continuous upon melting.
The continuity of the [13]C linewidth over a large temperature
range, above room temperature, that has previously been reported
(17) and is further detailed here does not reveal any change in
the vicinity of the α transition region (80-100°C). Major
changes are found in broad line proton NMR experiments, which
measure linewidths or second moments. (57) There is, however,
no discrepancy between these results since the α transition is a
property of the crystalline regions. The proton measurements

examine the complete sample, while the proton decoupled $^{13}C$
studies reported here are restricted to motions within the non-
crystalline regions.  These are unaffected by the α transition.

The proton spin-spin relaxation decay of unfractionated
polyethylene melts have been studied by Folland and Charlesby (58)
who interpreted their data in terms of the broad molecular weight
distribution.  The system was postulated to consist of high
molecular weight entangled molecules which coexisted with lower
molecular weight species.  Thus they argue for the presence of a
more mobile component as well as one which is subject to motional
constraints due to the entanglements.  Other processes could be
involved, but very likely for the same molecular reasons, such as
a distribution of correlation times (59) and incomplete motional
averaging of the dipolar interactions. (60)  It does not appear
necessary to require the existence of discretely different
structural entities, as has been argued. (61)

Further evidence for the direct influence of chain entanglements on the linewidth in the proton spectra for some completely
amorphous polymers has been demonstrated by Cohen-Addad and
collaborators. (60)(62)(63)  Spectral narrowing, as a consequence
of residual dipolar broadening, was observed in magic angle
spinning experiments, for polyisobutylene, polydimethyl siloxane
and cis 1,4 polybutadiene.  The significant result here is not
simply that the resonant linewidth was narrowed upon magic angle
spinning but that this effect was only observed over the concentration and molecular weight range where the chains were known to
be entangled.  For diluted systems, or for the pure polymers
whose molecular weights were lower than the critical value for
chain entanglement, no influence of the magic angle spinning was
observed.

In the amorphous state, therefore, at sufficiently high
molecular weights, polymer chains exhibit both liquid-like and
solid-like properties from the point of view of NMR measurements.
The liquid-like properties are manifested in the high frequency
segmental motions, and reflected in the spin-lattice relaxation
measurements.  The solid-like properties, as indicated by the
line broadening, are a result of residual dipolar couplings
caused by incomplete motional narrowing.  This latter effect is
removed as soon as the molecular basis for the entangled system
does not exist.

Conclusion

The results discussed above indicate that the further study
of the $^{13}C$ spin relaxation parameters possess the potential to
develop our understanding of the structure of the non-crystalline
regions of semicrystalline polymers.  Significant progress has
already been made in relating the spin-lattice relaxation
parameters with that of the pure melt.  The linewidths, or spin-
spin relaxation parameters, of semicrystalline polymers have been

shown to contain contributions from several major sources. The pathways and methods by which to sort these out have been set forth. Of particular importance is the influence of morphology, or crystallite arrangement, rather than the degree of crystallinity on the linewidths, which should be reflected in other properties. An inhomogeneous resonant line is typical of most semicrystalline polymers. A more detailed analysis of the line shapes that have been observed will be discussed elsewhere together with a more detailed discussion of the influence of field strength. (43) A clear picture has not as yet appeared on the influence of crystallinity and morphology on the nuclear Overhauser enhancement factor. More detailed work remains to be done in this area. However, preliminary results indicate that there is a major influence of the level of crystallinity.

Acknowledgement

This work was supported by the National Science Foundation under Grant No. DMR 76-21925.

Literature Cited

1.  L. Mandelkern, Morphology of Semicrystalline Polymers, in Characterization of Materials in Research: Ceramics and Polymers, Syracuse University Press, p. 369 (1975).

2.  E. W. Fischer, Prog. Colloid and Polymer Sci. 57, 149 (1975).

3.  J. H. Magill in Treatise in Material Science, V. 10, Part A, Academic Press, p. 1 (1977).

4.  L. Mandelkern, Crystallization of Polymers, McGraw-Hill (1964).

5.  L. Mandelkern, Acc. Chem. Res. 9, 81 (1976).

6.  S. Go, R. Prud'homme, R. Stein and L. Mandelkern, J. Polym. Sci., Polym. Phys. Ed. 12, 1185 (1974).

7.  L. Mandelkern, S. Go, D. Peiffer and R. S. Stein, J. Polym. Sci., Polym. Phys. Ed. 15, 1189 (1977).

8.  J. Maxfield and L. Mandelkern, Macromolecules 10, 1141 (1977)

9.  L. Mandelkern, J. Phys. Chem. 75, 3909 (1971).

10. V. D. Mochel, J. Macromol. Sci.-Revs. Macromol. Chem. C8, 289 (1972).

11. J. Schaefer, in Topics in Carbon-13 NMR Spectroscopy, Vol. 1, G. C. Levy, ed., Wiley-Interscience, New York, p. 149 (1974).

12. M. W. Duch and D. M. Grant, Macromolecules 3, 165 (1970).

13.  J. Schaefer, Macromolecules 5, 427 (1972).

14.  J. Schaefer, Macromolecules 6, 882 (1973).

15.  R. A. Komoroski and L. Mandelkern, J. Poly. Sci., Polym. Symp. C54, 201 (1976).

16.  R. A. Komoroski, J. Maxfield and L. Mandelkern, Macromolecules 10, 545 (1977).

17.  R. A. Komoroski, J. Maxfield, F. Sakaguchi and L. Mandelkern, Macromolecules 10, 550 (1977).

18.  J. Schaefer, E. O. Stejskal and R. Buchdahl, Macromolecules 8, 291 (1975).

19.  J. Schaefer and E. O. Stejskal, J. Amer. Chem. Soc. 98, 2035 (1976).

20.  H. A. Resing and W. B. Moniz, Macromolecules 8, 560 (1975).

21.  D. E. Axelson and L. Mandelkern, J. Polym. Sci., Polym. Phys. Ed. 16, 1135 (1978).

22.  D. E. Axelson and L. Mandelkern, to be published.

23.  G. C. Levy, I. R. Peat, R. Rosanske and S. Parks, J. Magn. Reson. 18, 205 (1975).

24.  D. Canet, G. C. Levy and I. R. Peat, J. Magn. Reson. 18, 199 (1975).

25.  J. Kowalewski, G. C. Levy, L. F. Johnson and L. Palmer, J. Magn. Reson. 26, 533 (1977).

26.  S. J. Opella, D. J. Nelson and O. Jardetzky, J. Chem. Phys. 64, 2533 (1976).

27.  J. G. Fatou and L. Mandelkern, J. Phys. Chem. 69, 71 (1965).

28.  E. Ergöz, J. G. Fatou and L. Mandelkern, Macromolecules 5, 147 (1974).

29.  S. Go, F. Kloos and L. Mandelkern, to be published.

30.  D. E. Axelson, G. C. Levy and L. Mandelkern, Macromolecules, in press.

31.  J. D. Hoffman, L. J. Frolen, G. S. Ross and J. I. Lauritzen, Jr., J. Res. Natl. Bur. Stand., Sect. A 79, 671 (1975).

32. F. Sakaguchi and L. Mandelkern, unpublished observations.

33. J. Maxfield and L. Mandelkern, J. Polym. Sci., Polym. Phys. Ed., in press.

34. Y. Inoue, A. Nishioka, and R. Chujo, Makromol. Chem. 168, 163 (1973).

35. J. Schaefer and D. F. S. Natusch, Macromolecules 5, 416 (1972).

36. J. J. Lindberg, I. Sirén, E. Rahkamaa and P. Tormala, Die Angewandte Makromolekulare Chemie 50, 187 (1976).

37. D. W. McCall, D. C. Douglass and E. W. Anderson, J. Polym. Sci. 59, 301 (1962).

38. C. L. Beatty and M. F. Froix, Polym. Prepr. Am. Chem. Soc. Div. Polym. Chem. 16, 628 (1975).

39. T. M. Connor and A. Hartland, J. Polym. Sci., Polym. Phys. Ed. 7, 1005 (1969).

40. J. A. Pople, W. G. Schneider and A. Bernstein, High Resolution Nuclear Magnetic Resonance, McGraw-Hill, p. 80 (1959).

41. J. K. Becconsall, P. A. Curnuck and M. C. McIvor, Appl. Spec. Rev. 4, 307 (1971).

42. C. P. Poole and H. A. Farrach, Relaxation in Magnetic Resonance, Academic Press (1971).

43. D. E. Axelson, R. A. Komoroski and L. Mandelkern, to be published.

44. N. Bloembergen, E. M. Purcell and R. V. Pound, Phys. Rev. 73, 679 (1948).

45. J. Schaefer, J. Magn. Resonance 6, 670 (1972).

46. S. Kaufman, W. P. Slichter and D. D. Davis, J. Polym. Sci., Polym. Phys. Ed. 9, 829 (1971).

47. J. Schaefer, S. H. Chin and S. I. Weissman, Macromolecules 5, 798 (1972).

48. F. Bueche, Physical Properties of Polymers, Interscience, p. 65ff (1962).

49. F. Bueche, J. Chem. Phys. 20, 1959 (1952); ibid., 25, 599 (1956).

50.  W. W. Graessley, Adv. Polym. Sci. 16, 1 (1974).

51.  J. D. Ferry, Viscoelastic Properties of Polymers, 2nd
     edition, Wiley (1970).

52.  G. C. Berry and T. G. Fox, Adv. Polym. Sci. 5, 261 (1968).

53.  R. S. Porter and J. F. Johnson, J. Appl. Polym. Sci. 3, 194
     (1960).

54.  R. S. Porter and J. F. Johnson, Trans. Soc. Rheol. 6, 107
     (1962); Soc. Plast. Eng. Trans. 3, 18 (1963).

55.  H. Markovitz, T. G. Fox and J. D. Ferry, J. Phys. Chem. 66,
     1567 (1962).

56.  T. P. Yin, S. E. Lovell, and J. D. Ferry, J. Phys. Chem. 65,
     534 (1961).

57.  H. G. Olf and A. Peterlin, J. Polym. Sci. A-2 8, 753, 771
     (1970).

58.  R. Folland and A. Charlesby, J. Polym. Sci., Polymer Lett.
     Ed. 16, 339 (1978).

59.  F. Horii, R. Kitamaru and T. Suzuki, J. Polym. Sci., Polym.
     Lett Ed. 15, 65 (1977).

60.  J. P. Cohen-Addad, M. Domard and J. Herz, J. Chem. Phys. 68,
     1194 (1978).

61.  W. L. F. Gölz, H. G. Zachmann, Kolloid-Z.Z. Polym. 247, 814
     (1971).

62.  J. P. Cohen-Addad and C. Roby, J. Chem. Phys. 63, 3095
     (1975).

63.  J. P. Cohen-Addad and J. P. Faure, J. Chem. Phys. 61, 1571
     (1974).

## Discussion

J. Guillet, University of Toronto, Ontario:  Just one comment about your glass transition temperatures.  Surely because
the frequency with which you are working is so high, the glass
transition temperature will occur at a much lower temperature
rather than at a much higher one.  Would this not be correct?

D. Axelson:  The glass temperature usually increases with
increasing frequency.  However, in the present problem our
conclusions are based on the correlation time, which is a
frequency-independent quantity.

J. Guillet:  What is the essential frequency of your measurement?  Presumably it is not the frequency of the radiation.

D. Axelson: These spectra were obtained at 67.9 MHz, but that's not the problem. We can measure correlation times regardless of the frequency. The correlation time at the glass temperature is very long. From a measurement of the correlation time we should be able to tell whether it is a true glass. In all these cases the correlation times are six to nine orders of magnitude lower than can possibly exist in a glass. For this reason I think the correlation between the NMR measurement and dielectric relaxation and dynamic mechanical do not relate one to one because of the frequency effects in the other measurements.

J. Guillet: But you would expect a frequency effect in this one as well, if only the frequency of the motion itself.

D. Axelson: The relaxation parameters are frequency dependent but not the correlation time.

J. C. Randall, Phillips Petroleum, Oklahoma: I was interested in your linewidth curves vs temperature in which you essentially had the melting point curves. Did you do any freezing point curves?

D. Axelson: These were obtained for branched polyethylene and were similar to the melting point curves. The conditions were such that either direction gave identical linewidths.

J. C. Randall: Yes, I could see that in the low density polyethylene. It would be interesting to compare a system of spherulites vs known morphology for essentially the same crystallinity. What result would those conditions give?

D. Axelson: This would be a very interesting experiment to carry out but technically somewhat difficult for us at present. We have plans to carry out such measurements in the near future.

C. J. Carman, B. F. Goodrich, Ohio: Since Tg is a zero frequency measurement and since the NMR experiment is at a higher frequency, I think Tg would go to a higher value. In other words, your apparent Tg with an NMR measurement would be higher than a Tg as measured with a zero frequency measurement (DSC). Therefore I don't think the numbers you presented are too surprising in view of the fact that you are at a higher frequency. Your Tc should be a pseudofunction of Tg at higher temperatures than of a Tg measured by a thermal measurement.

D. Axelson: Based on correlation time measurements the upper limit Tg's that we report are frequency-independent. The results would only be surprising to those who argue for a much higher Tg for the polyethylenes.

A. Jones, Clark University, Massachusetts: Is the dominant effect on $T_2$ morphological and the dynamics of low frequency motions somewhat difficult to extract from $T_2$?

D. Axelson: For the polyethylenes, at least, there is a major effect of morphology on linewidth. This is going to make more difficult a detailed description of the dynamics of the low frequency motion relative to a completely amorphous polymer.

J. Guillet: I thank Dr. Carman for pointing out my error caused by the early hour of the morning, I think. It is quite correct that Tg should be higher and, as I recall from the work on dielectric and other measurements of the glass transition, a

ten degree rise in the glass transition might be expected for
every log hertz. This would correspond to a glass transition at
about ten to the fourth or fifth cycles per second, which sounds
about the right range for the NMR. Fifty degrees' difference
would be expected, ten degrees for each order of magnitude change
in the frequency. It could correlate quite well with the glass
transition temperature.

D. Axelson: As we have already pointed out, the correlation
time is frequency-independent. The longest correlation time that
we have measured is about $10^{-6}$ s. Whether the results correlate
well with the glass temperature depends on the value one accepts
for linear and branched polyethylene. Those values have been a
controversial matter.

J. Guillet: It seems to me interest should be focused on
only one kind of correlation time and that is the one that relates
to the motions of long segments in the polymer. Obviously, the
glass transitions are not going to be affected by rotations of
methyl groups and phenyl groups. The motion that really relates
to the glass transition is whether or not a complete reordering
of segments of ten- to fifty-monomer units takes place. This
correlates with the glass transition.

D. Axelson: Carbon-13 NMR allows for the measurement of the
average correlation time for each individual carbon atom. For
the glass temperature problem we are obviously only concerned
with the correlation time of the backbone carbons.

C. J. Carman: Earlier in your talk you showed the carbon
$T_1$ data and NOEF for partially crystalline and amorphous poly-
isoprenes. Was this a natural rubber which had been allowed to
crystallize to different degrees or was this a synthetic rubber?

D. Axelson: The sample studied was a synthetic cis-
polyisoprene. Its cis-1,4 content was greater than 99%.

C. J. Carman: How was the crystallinity controlled and how
was the crystallinity ascertained? As I recall the data, the
$T_1$'s were not affected; neither were the NOEF's. In examining
the amorphous region how can one be certain the crystalline
region is participating in the data anyway?

D. Axelson: Detailed answers to these questions can be
found in Macromolecules 10, 545 (1977); ibid., 10, 55 (1977).

P. Sipos, Dupont, Ontario: In the case of polyethylene
what was the origin of the sample? Because it makes a difference
as far as being catalytic or free radical.

D. Axelson: The low density (branched) polyethylenes were
free radical initiated; the linear polymers were derived from
commercial sources and purified and characterized as described
(Macromolecules 10, 550(1977)).

RECEIVED March 13, 1979.

# Use of Carbon-13 NMR Analysis in the Characterization of Alternating Copolymers Prepared by Chemical Modifications of 1,4-Polydienes

R. LACAS, G. MAURICE, and J. PRUD'HOMME

Department of Chemistry, University of Montreal, Montreal, Quebec, Canada H3C 3V1

Chemical modifications of unsaturated polymers, especially polydienes, have been investigated for many years in order to derive novel polymers having specific physical and mechanical properties (1). More recently, interest in elastomers with higher stability to oxidative degradation has intensified research into hydrogenation of polydienes (2,3), including polydiene moieties in thermoplastic block and graft copolymers (4). Another feature of chemical modifications of polydienes is their ability to provide a route for preparing novel polymers of particular structures such as head-to-head polymers or alternating copolymers which cannot be prepared by conventional polymerization or copolymerization processes. For example, equimolar alternating copolymers of ethylene with propylene can be obtained by hydrogenating 1,4-polyisoprene (3,5), while head-to-head polypropylene can be obtained by hydrogenating 1,4-poly(2,3-dimethyl-1,3-butadiene) (1,4-polydimethylbutadiene) (5). Hydrohalogenation reactions carried out on 1,4-polydienes may also result in equimolar alternating copolymers of well defined structures (6).

In the present paper we wish to report both $^1H$ and $^{13}C$ NMR studies of such alternating copolymers obtained by hydrogenation and hydrohalogenation reactions carried out on 1,4-polydienes prepared with butyllithium in nonpolar solvents. Available for this work were hydrogenated 1,4-polyisoprene and 1,4-polydimethylbutadiene, hydrochlorinated 1,4-polyisoprene and 1,4-polydimethylbutadiene, and hydrobrominated 1,4-polyisoprene. Unlike 1,4-polyisoprene, 1,4-polydimethylbutadiene has symmetrically tetrasubstituted double bonds in its backbone. This situation gives rise to the possibility of threo and erythro diastereoisomerism in the repeating units of the saturated materials. Also of interest, is the possibility of head-to-head and tail-to-tail additions of hydrogen halide in 1,4-polydimethylbutadiene. Both these features were investigated on the basis of the $^{13}C$ chemical shift substituent effects derived from spectra measured on suitable model compounds, including hydrohalogenated 1,4-polyisoprene.

0-8412-0505-1/79/47-103-215$05.25/0
© 1979 American Chemical Society

## Preparation and 220MHz $^1$H NMR Characterization of the Materials

Starting Materials. The polydienes submitted to the chemical modifications were polyisoprene and polydimethylbutadiene samples prepared by anionic polymerization using sec-butyl lithium as initiator and hydrocarbons as solvents (5,6). The polyisoprene sample was prepared at 25°C using benzene as solvent (6). Its microstructure determined by $^1$H NMR spectroscopy was 71% cis-1,4, 22% trans-1,4, and 7% 3,4. Its number average molecular weight was 8.6 x $10^4$. The polydimethybutadiene sample was prepared at 60°C using cyclohexane as solvent (5). Its microstructure determined by $^1$H NMR spectroscopy was 74% trans-1,4, 23% cis-1,4, and 3% 1,2. Its number average molecular weight was 4.3 x $10^4$.

Hydrogenation Reactions. The hydrogenation reactions were carried out on 0.5% polymer solutions in cyclohexane by using coordination catalysts made by the reaction of triethylaluminum with the cobalt (II) salt of 4-cyclohexylbutanoic acid. For that purpose, catalyst solutions having an aluminum/cobalt molar ratio of 4 were prepared under a nitrogen atmosphere by slowly adding a molar solution of triethylaluminum to a 2 x $10^{-2}$ M solution of the cobalt salt, both in cyclohexane. The catalyst solutions were added to the polydiene solutions under a nitrogen atmosphere following which hydrogen was bubbled through the solutions at a constant pressure of 4 atm for a period of 2h, at 50°C. Under these experimental conditions, quantitative hydrogenation of the 1,4-polyisoprene sample was obtained using 5 mol % of catalyst based on unsaturated monomer units, whereas nearly quantitative hydrogenation of the 1,4-polydimethylbutadiene sample required 30 mol % of catalyst. Completion of the reactions is demonstrated in Figures 1 and 2 where the upfield regions of the 220 MHz $^1$H NMR spectra measured before and after hydrogenation are presented for the 1,4-polyisoprene and the 1,4-polydimethylbutadiene samples, respectively.

From Figure 1, one can see that no trace of residual unsaturated units is detectable in the $^1$H spectrum of the hydrogenated 1,4-polyisoprene sample. On the other hand, the spectrum exhibits a well resolved methyl doublet (J = 6.5 Hz) centered at 0.92 ppm. From Figure 2, one can see that the resonances of the unsaturated units are not completely absent in the $^1$H spectrum of the hydrogenated 1,4-polydimethylbutadiene sample. However relative intensity measurements made on the spectrum indicate a degree of saturation higher than 98% for this polymer. Interestingly, the methyl resonance of the hydrogenated 1,4-polydimethylbutadiene sample appears as two doublets centered at 0.90 ppm indicating that two different configurations can be distinguished for this polymer. In fact, assuming the same coupling constant, J = 6.5 Hz, as that observed in the spectrum of hydrogenated 1,4-polyisoprene, one can interpret the methyl resonances of hydrogenated 1,4-polydimethylbutadiene as the

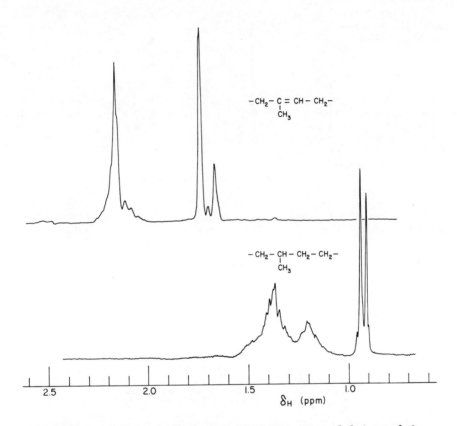

*Figure 1.   Upfield region of the 220 MHz ¹H NMR spectra made before and after the hydrogenation of the 1,4-polyisoprene sample. Chlorbenzene solutions at 100°C with TMS as reference.*

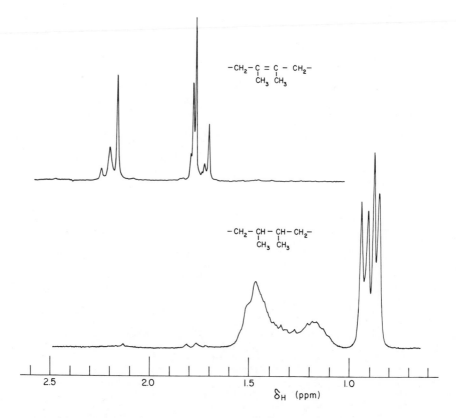

Figure 2.   Upfield region of the 220 MHz ¹H NMR spectra made before and after
the hydrogenation of the 1,4-polydimethylbutadiene sample.  Chlorobenzene solutions at 100°C with TMS as reference.

juxtaposition of two doublets with a chemical shift difference of
13.5 Hz.  One of these doublets would correspond to the threo
configuration and the other to the erythro configuration of the
saturated repeating units.

  Hydrohalogenation Reactions.  Both the hydrochlorination and
hydrobromination of the 1,4-polyisoprene sample were conducted at
25°C on 1% polymer solutions in toluene.  The solutions were first
purged with dry nitrogen after which the dry hydrogen halide was
bubbled through the reaction system at a pressure slightly above
atmospheric for a period of 24 h.  Figure 3 shows the 220 MHz
[1]H NMR spectra measured on the two hydrohalogenated products.  In
either [1]H spectrum quantitative hydrohalogenation is indicated by
the loss of the isoprene unit methylene resonances at 2.2 ppm.
On the other hand, the fact that the methyl resonances appear as
sharp singlets at 1.53 ppm for the hydrochlorinated product and
at 1.69 ppm for the hydrobrominated product indicates the
exclusiveness of Markownikoff's rule for the addition of either
hydrogen halide to the repeating units of 1,4-polyisoprene.

  The hydrochlorination of the 1,4-polydimethylbutadiene sample
was conducted using the same procedure as for 1,4-polyisoprene,
except that it was necessary to increase the hydrogen chloride
pressure to 4 atm in order to obtain a quantitative saturation of
this polymer.  Figure 4 shows the 220 MHz [1]H NMR spectrum measured
on the product.  Owing to the possible superposition of the [1]H
resonances of the hydrochlorinated units with those of the
unsaturated units (see Figure 2), estimation of the degree of
conversion from the [1]H spectrum alone is difficult.  However,
elemental analysis of the product indicated a degree of hydrochlorination close to 99%.  From Figure 4, one can see that the
two methyl resonances expected for hydrochlorinated 1,4-
polydimethylbutadiene appear as pairs of complex signals centered
at 1.1 and 1.5 ppm, respectively.  The signals centered at 1.1 ppm
arise from the methyl groups on the tertiary carbons.  As for
hydrogenated 1,4-polydimethylbutadiene, it is reasonable to
interpret these signals as the juxtaposition of two doublets
corresponding to threo and erythro configurations of the repeating
units although, in the present case, the resolution of the two
doublets is not as good as for the hydrogenated polymer.  The
other two signals centered at 1.5 ppm arise from the methyl groups
on the quaternary carbons.  Apparent in either signal is a fine
structure which suggests sensitivity to the position of the
chlorine atoms in the neighbouring units.  The same effect may
also explain the lack of resolution observed for the methyl
resonances centered at 1.1 ppm.  This point is clarified in the
next section where the same material is characterized by [13]C NMR
spectroscopy.

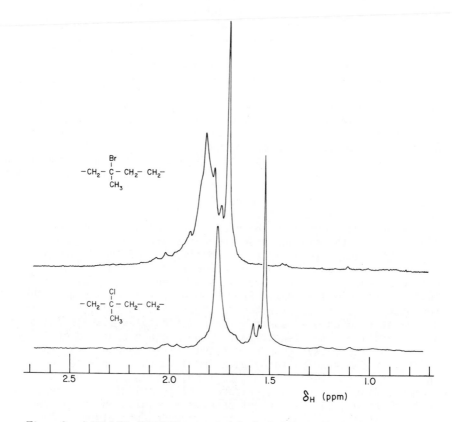

*Figure 3.    A 220-MHz $^1$H NMR spectra of the hydrohalogenated 1,4-polyisoprene samples.  Chlorobenzene solutions at 100° C with TMS as reference.*

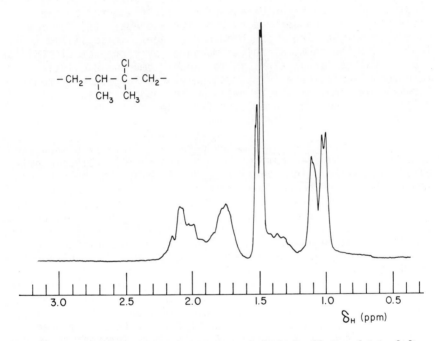

*Figure 4. A 220-MHz ¹H NMR spectrum of the hydrochlorinated 1,4-polydimethylbutadiene sample. Chlorobenzene solution at 100°C with TMS as reference.*

## $^{13}$C NMR Characterization of the Materials

Hydrogenated 1,4-polyisoprene. Figure 5 shows the 22.6 MHz $^{13}$C NMR spectrum of the hydrogenated 1,4-polyisoprene sample. The spectrum is identical to those already reported in the literature for hydrogenated natural rubber (7) as well as for hydrogenated synthetic cis-1,4-polyisoprene (8). It consists of four main signals which are identified in Figure 5. The spectrum also exhibits much weaker signals which arise from the hydrogenated 3,4 units present in the polymer. As shown in Table I, the chemical shifts of the four main resonances observed in the spectrum of Figure 5 can be satisfactorily predicted by the empirical equation derived by Lindeman and Adams (9) for low molecular weight linear and branched alkanes. Also listed in Table I are the chemical shifts measured by Carman et al. (8) for the equivalent carbon atoms in the model compound 2,6,10,14-tetramethylpentadecane. Interestingly, the largest difference between the chemical shifts measured for hydrogenated 1,4-polyisoprene and those predicted from the model compound resonances does not exceed 0.2 ppm.

Hydrogenated 1,4-polydimethylbutadiene. Figure 6 shows the 22.6 MHz $^{13}$C NMR spectrum of the hydrogenated 1,4-polydimethylbutadiene sample. The identification of methyl, methylene, and methine carbon resonances was made by off-resonance decoupling experiments. Two distinct signals occur for each type of carbon suggesting discrimination between threo and erythro configurations for the repeating units. Such a discrimination was also apparent in the 220 MHz $^1$H NMR spectrum (Figure 2) which exhibited two distinct methyl doublets centered at 0.90 ppm. This interpretation is substantiated by comparing the $^{13}$C NMR spectrum of Figure 6 with that measured by Lindeman and Adams (9) for a mixture of the two diastereoisomers of 3,4-dimethylhexane. The latter spectrum (not shown) also exhibits two distinct resonances for all carbons but the terminal methyl carbons which are separated from the tertiary carbons by one methylene unit. More interesting is the fact that very similar chemical shift differences are observed for each pair of signals in either case: 2 ppm for both the methyl and methylene resonances and 1 ppm for the methine resonances.

Hydrogenated 1,4-polydimethylbutadiene may be considered as either an alternating copolymer of ethylene and 2-butene or as a head-to-head polypropylene. Such alternating structures of ethylene and 2-butene monomer units can be obtained directly by copolymerizing the two olefins with Ziegler-Natta catalysts. For instance, Natta et al. (10) reported that cis-2-butene can be copolymerized with ethylene using $VCl_4/AlR_3$ catalyst systems to yield crystalline alternating copolymers of the erythro-diisotactic structure. The $^{13}$C NMR spectrum of this latter copolymer has been reported by Zambelli et al. (11). As shown

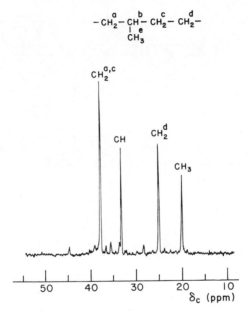

*Figure 5.   Proton noise-decoupled 22.6-MHz C-13 NMR spectrum of the hydrogenated 1,4-polyisoprene sample. Perdeuteriobenzene solution at 25°C with TMS as internal reference. Approximately 5000 pulses with an acquisition time of 0.7 sec and a flip angle of 30°.*

Table I.

Observed and Predicted $^{13}$C Chemical Shifts*

for Hydrogenated 1,4-Polyisoprene

| Carbon** | Chemical Shifts (ppm) | | |
|---|---|---|---|
| | Observed | Predicted | |
| | | Lindeman-Adams | Model compound*** |
| CH$_3$ | 20.1 | 19.6 | 20.2 |
| CH$_2^d$ | 25.1 | 24.6 | 25.3 |
| CH | 33.4 | 32.5 | 33.3 |
| CH$_2^{a,c}$ | 38.1 | 36.9 | 38.0 |

  * All chemical shifts are in ppm downfield from TMS.
 ** Identified in Figure 5.
*** Measured by Carman et al. ($\underline{8}$) on 2,6,10,14-tetramethylpenta-
    decane.

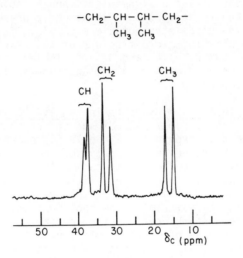

*Figure 6. Proton noise-decoupled 22.6-MHz C-13 NMR spectrum of the hydrogenated 1,4-polydimethylbutadiene sample. Perdeuteriobenzene solution at 25°C with TMS as internal reference. Approximately 9000 pulses with an acquisition time of 0.7 sec and a flip angle of 30°.*

in Table II, comparison of the chemical shifts measured by
Zambelli et al. with those observed for the hydrogenated 1,4-
polydimethylbutadiene sample allows the assignments of the erythro
signals in the latter. The results are indicated in Table II
beside the chemical shifts of the observed signals. One can see
that the largest difference in chemical shift between the two
sets of matched signals does not exceed 0.2 ppm. Note that an
additional methylene carbon resonance was observed at 30.3 ppm
(downfield from TMS) in the spectrum of the copolymer studied
by Zambelli et al. They attributed this resonance to internal
methylene carbons in sequences of more than one ethylene unit
because the copolymer contained more than 50 mol % of ethylene
units.

On the basis of the assignments given in Table II, one may
conclude that the threo structure is slightly predominant in the
hydrogenated 1,4-polydimethylbutadiene sample. This result taken
in conjunction with the microstructure of the starting material
(74% trans-1,4, 23% cis-1,4, and 3% 1,2) indicates a nonstereo-
specific addition of hydrogen to the unsaturated 1,4 units of
polydimethylbutadiene.

Hydrochlorinated and Hydrobrominated 1,4-Polyisoprene.
Figure 7 shows the 22.6 MHz $^{13}$C NMR spectra of both the hydrochlorinated and hydrobrominated 1,4-polyisoprene samples. Like
the spectrum of hydrogenated 1,4-polyisoprene, either spectrum
in Figure 7 exhibits four major signals only, the assigments of
which are indicated in the Figure. Either spectrum also shows
much weaker signals which arise from the hydrohalogenated 3,4
units present in the material. The fact that only four major
carbon resonances are observed in the spectra of Figure 7 results
from the exclusiveness of Markownikoff's rule for the hydrogen
halide addition to 1,4-polyisoprene, a characteristic which was
previously evidenced from the 220 MHz $^1$H NMR spectra (Figure 3).
This situation is of interest, since such halogen atoms regularly
introduced in a polyolefinic chain, particularly bromine atoms,
might be used as intermediates for subsequent chemical
modifications based upon substitution reactions.

Comparing the $^{13}$C NMR spectrum data of the hydrohalogenated
1,4-polyisoprene samples with those of the hydrogenated sample
allows one to evaluate the shielding contributions produced by
the halogen substituents in the surrounding $\alpha$, $\beta$, and $\gamma$ positions
in the polymer chains. The results of this comparison are
summarized in Table III (upper part) where the substituent effects
upon replacement of the hydrogen atom on the tertiary carbon by
either a chlorine or a bromine atom are listed for the 1,4-
polyisoprene derivatives. One can see that the differences
between the chlorine and the bromine substituent effects range
from 1.0 ppm for the $\gamma$ effect to 1.6 ppm for the $\alpha$ effect.
Interestingly, the set of shift parameters in the upper part of
Table III is very close to that which one can compute from

Table II.

<u>$^{13}$C Chemical Shifts* of Hydrogenated 1,4-Polydimethylbutadiene</u>

<u>Compared to those of cis-2-Butene-Ethylene</u>

<u>Alternating Erythrodiisotactic Copolymer</u>

| Carbon | Hydrogenated 1,4-Polydimethylbutadiene | cis-2-Butene-Ethylene Alternating Copolymer** |
|--------|------|------|
| CH$_3$ | 14.7 (t)*** <br> 17.0 (e) | 16.8 |
| CH$_2$ | 31.4 (e) <br> 33.5 (t) | 30.3 <br> 31.5 |
| CH | 37.5 (t) <br> 38.5 (e) | 38.6 |

* All chemical shifts are in ppm downfield from TMS.
** Data from Zambelli et al. (<u>11</u>). Original chemical shifts were converted downfield from HMDS to downfield from TMS using δ(HMDS) = 2.0 ppm.
*** Assignment of <u>threo</u> (t) and <u>erythro</u> (e) signals.

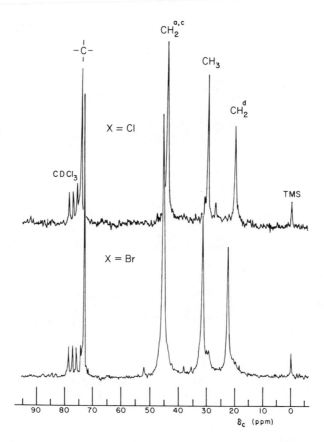

Figure 7.    Proton noise-decoupled 22.6-MHz C-13 NMR spectra of the hydrohalogenated 1,4-polyisoprene samples.   Deuteriochloroform solutions at 25°C
with TMS as internal reference.   Approximately 5000 pulses with an acquisition
time of 0.7 sec and a flip angle of 30°.

Table III.

Substituent Effects upon Replacement of Hydrogen by Halogen

in

$$\begin{array}{ccccc} & & X & & \\ \beta' & \alpha| & \beta' & \gamma & \\ -CH_2 & - & C & - & CH_2 & - & CH_2 & - \\ & & \beta| & & \\ & & CH_3 & & \end{array}$$

| X | $\alpha$ | $\beta$ | $\beta'$ | $\gamma$ |
|----|----------|---------|----------|----------|
| Cl | + 40.9 | + 9.7 | + 6.1 | - 2.4 |
| Br | + 39.3 | + 11.3 | + 7.2 | - 1.4 |

---

in

$$\begin{array}{ccccc} & & X & & \\ \beta & \alpha| & \beta' & \gamma & \\ CH_3 & - & C & - & CH_2 & - & CH_3 \\ & & \beta| & & \\ & & CH_3 & & \end{array}$$

| X | $\alpha$ | $\beta$ | $\beta'$ | $\gamma$ |
|----|----------|---------|----------|----------|
| Cl | + 40.8 | + 9.4 | + 7.0 | - 2.3 |
| Br | + 37.9 | + 11.6 | + 8.4 | - 1.1 |

standard source spectra for 2-methylbutane and its 2-chloro and 2-bromo derivatives. For the sake of comparison, this latter set of shift parameters is listed in the lower part of Table III.

From Table III, one can see that a common feature in both sets of shift parameters (that for 1,4-polyisoprene derivatives and that for 2-methylbutane derivatives) is a higher β effect for a methyl carbon (denoted β in Table III) than for a methylene carbon (denoted β' in Table III). The difference between β and β' effects is close to 4 ppm for the polymers whereas it is close to 3 ppm for the model compounds. On the other hand, one can see that there is no significant difference in the γ effects whether the carbon under consideration is a methyl or methylene carbon. The empirical shift parameters calculated from the $^{13}$C NMR data of the 1,4-polyisoprene derivatives will provide, hereafter, a basis for assigning the carbon resonances observed in the spectrum of hydrochlorinated 1,4-polydimethylbutadiene, which like hydrochlorinated 1,4-polyisoprene has quaternary carbons substituted by one chlorine atom.

   Hydrochlorinated 1,4-polydimethylbutadiene. Figure 8 shows the 22.6 MHz $^{13}$C NMR spectrum of the hydrochlorinated 1,4-polydimethylbutadiene sample. Identification of methyl, methylene, methine, and quaternary carbon resonances was made by off-resonance decoupling experiments. It can be seen that all resonances but the methyl resonance at 15.1 ppm, and the methine resonance at 44.3 ppm, occur as pairs of signals. On the other hand, the off-resonance spectrum (not shown) revealed the presence of a methylene resonance overlapping the pair of methyl signals centered at 27.0 ppm. In fact, the spectrum in Figure 8 contains four distinct methylene resonances centered at 27.1, 31.1, 37.0, and 41.4 ppm. This feature arises because there is no selective induction mechanism for controlling the position of the chlorine atom in the addition of hydrogen chloride to symmetrically substituted double bonds such as those in 1,4-polydimethylbutadiene. This situation leads to the possibility of head-to-tail, head-to-head, and tail-to-tail arrangements of two consecutive hydrochlorinated 1,4-units. These three arrangements are depicted in Figure 8. They show four chemically nonequivalent methylene carbons which are denoted as $CH_2^c$, $CH_2^d$, $CH_2^{c'}$, and $CH_2^{d'}$.

   The chemical shifts of the four methylene carbons depicted in Figure 8 have been calculated using the empirical shift parameters derived from the preceding study on hydrochlorinated 1,4-polyisoprene. They are listed in Table IV together with the chemical shifts predicted for the other carbon atoms in hydrochlorinated 1,4-polydimethylbutadiene. All the calculations were based on the mean values of the chemical shifts observed for the threo and erythro diastereoisomers in hydrogenated 1,4-polydimethylbutadiene, i.e. 15.9 ppm for $CH_3$, 32.5 ppm for $CH_2$, and 38.0 ppm for CH. Also listed in Table IV are the experimental chemical shifts. The complete assignments given in both Table IV and Figure 8 were made by fitting the observed methyl and methylene

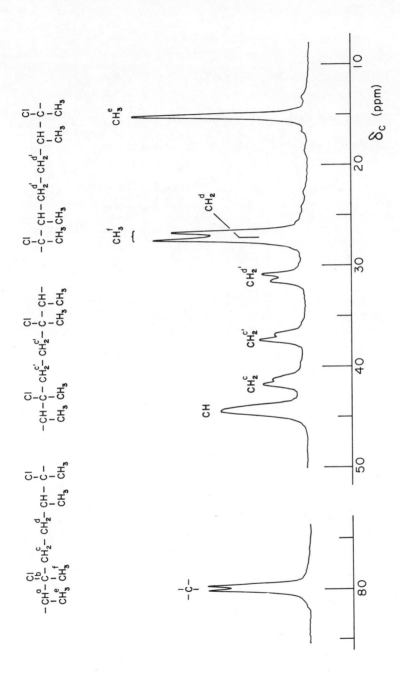

*Figure 8. Proton noise-decoupled 22.6-MHz C-13 NMR spectrum of the hydrochlorinated 1,4-polydimethylbutadiene sample. $CD_2Cl_2$ solution at 25°C with TMS as internal reference. Approximately 15000 pulses with an acquisition time of 1.6 sec and a flip angle of 90°.*

Table IV.

Observed and Predicted $^{13}$C Chemical Shifts* for

Hydrochlorinated 1,4-Polydimethylbutadiene

| Carbon** | Chemical Shifts (ppm) | |
|---|---|---|
| | Observed | Predicted |
| $CH_3^e$ | 15.1 | 13.4 |
| $CH_3^f$ | 26.6 27.3 | 25.5 |
| $CH_2^d$ | 27.1 | 27.6 |
| $CH_2^{d'}$ | 30.8 31.5 | 30.1 |
| $CH_2^{c'}$ | 36.7 37.3 | 36.2 |
| $CH_2^c$ | 41.2 41.7 | 38.6 |
| CH | 44.3 | 44.1 |
| $-\overset{\mid}{\underset{\mid}{C}}-$ | 79.6 80.1 | 78.9 |

* All chemical shifts are in ppm downfield from TMS.
** Identified in Figure 8.

carbon chemical shifts to the predicted shifts. From Table IV, one can see that a fairly good fitting was obtained for all resonances except the methyl resonance observed at 15.1 ppm and the methylene resonance centered at 41.5 ppm. The difference between the predicted and observed shifts is 1.7 ppm for the former and 3 ppm for the latter.

Several carbon resonances occur as pairs of signals in the spectrum of Figure 8. As for hydrogenated 1,4-polydimethyl-butadiene, this may be attributed to the presence of repeating units having threo and erythro configurations in hydrochlorinated 1,4-polydimethylbutadiene. Note that a similar splitting into pairs of signals was also observed for the two methyl resonances in the 220 MHz $^1$H NMR spectrum of the same material (Figure 4). In contrast with the $^1$H spectrum, only the methyl carbon on the quaternary carbon gives rise to a pair of signals in the $^{13}$C spectrum. The other methyl resonance which arises from the methyl group on the tertiary carbon appears as a single signal at 15.1 ppm in the $^{13}$C spectrum. The fact that this latter $^{13}$C resonance does not exhibit threo and erythro configurational sensitivity might be explained by a fortuitous magnetic equivalence due to steric interactions of the methyl group with the three $\gamma$ substituents on the quaternary carbon.

Quantitative analysis of the 22.6 MHz $^{13}$C NMR spectrum in Figure 8 yields the following information concerning the microstructure of the hydrochlorinated 1,4-polydimethylbutadiene sample. First, the population of one of the diastereoisomers, threo or erythro, is slightly predominant in the material. This is substantiated by the relative intensities of the signals which occur as pairs in the spectrum of Figure 8. This is also supported by the relative intensities of the two methyl doublets observed at 1.1 ppm in the 220 MHz $^1$H NMR spectrum of Figure 4. Second, a random placement of the chlorine atoms occured in the course of the hydrochlorination reaction. This latter result is based on the nearly equal intensities measured for the three methylene carbon resonances directly observable in the spectrum of Figure 8. Note that a fourth methylene carbon resonance was detected at 27.1 ppm by off-resonance decoupling experiments. As shown previously, this methylene resonance at 27.1 ppm and that observed at 41.4 ppm each arise from one of the two chemically nonequivalent methylene carbons in head-to-tail arrangements of two consecutive hydrochorinated units. On the other hand, the methylene resonances observed at 31.1 and 37.0 ppm each arise from the two chemically equivalent methylene carbons in tail-to-tail and head-to-head arrangements, respectively. Therefore the equal intensities measured for the three methylene resonances observed at 31.1, 37.0, and 41.4 ppm indicate the 1:2:1 proportions of head-to-head, head-to-tail, and tail-to-tail arrangements expected for a random placement of the chlorine atoms in the course of the hydrochlorination reaction.

## Acknowledgments

This work was supported by the National Research Council of Canada and the Quebec Ministry of Education. The authors wish to thank the Canadian 220 MHz NMR Centre for making the proton NMR measurements and Mr. R. Mayer for          the carbon NMR spectra.

## Literature Cited

1.  E.M. Fettes, "Chemical Reactions of Polymers", Interscience
    Publ., New York, 1964, Chap. II.
2.  J.C. Falk, J. Polym. Sci., Part A-1, 9, 2617 (1971).
3.  L.A. Mango, and R.W. Lenz, Makromol. Chem., 163, 13 (1973).
4.  R.C. Jones, U.S. Pat. 3,431,323 (March 4, 1969).
5.  D. Khlok, Y. Deslandes, and J. Prud'homme, Macromolecules,
    9, 809 (1976).
6.  A. Tran and J. Prud'homme, Macromolecules, 10, 149 (1977).
7.  Y. Tanaka, H. Sato, A. Ogura, and I. Nagoya, J. Polym. Sci.,
    Polym. Chem. Ed., 14, 73 (1976).
8.  C.J. Carman, A.R. Tarpley, and J.H. Goldstein, Macromolecules,
    6, 719 (1973).
9.  L.P. Lindeman, and J.Q. Adams, Anal. Chem., 43, 1245 (1971).
10. G. Natta, G. Dall'Asta, G. Mazzanti, I. Pasquon,
    A. Valvassori, and A. Zambelli, J. Am. Chem. Soc., 83, 3343
    (1961).
11. A. Zambelli, G. Gatti, S. Sacchi, W.O. Crain, and
    J.D. Roberts, Macromolecules, 4, 475 (1971).

## Discussion

J. Randall, Phillips Petroleum, Oklahoma:  I was interested in the β and β' effects for the halogen substituents.  Have you looked at the Grant and Paul approach in which they introduced corrective terms?  The change from a tertiary to a quaternary carbon would cause conformational changes.  Grant and Paul faced the same problem when looking at methyl group substitutions.  I was wondering if the differences in the β and β' terms might disappear if you did this.  Any conformational contribution might disappear. It would be interesting to see how close the two β terms come.  I found this to work on the substituent effect for an aromatic ring. I introduced corrective terms and got the same β or α.  It may not happen here because the conformational change may not be predicted by the simple alkyl shifts.

J. Prud'homme:  We did not make the corrections.  This is a good suggestion.

G. Babbitt, Allied Chemical Corp., N.J.:  How do you know that a methylene resonance was under the methyl signals in the $^{13}C$ spectrum of hydrochlorinated 1,4-polydimethylbutadiene?

J. Prud'homme: Through the use of an off-resonance measurement. The methyl signals appeared as two juxtaposed quartets in the middle of which we could see a methylene resonance.

G. Babbitt. In other words the methyl signals opened by going to quartets and revealed the triplet underneath.

J. Prud'homme: Exactly.

G. Babbitt: In the same spectrum, you showed pairs of methylene signals and attributed the doubling to erythro and threo structures of the H-Cl units. Three units were pictured: an ethylene unit in the center with H-Cl containing units on either end. It is possible that both ends can be erythro, or both ends can be threo, or be mixed. Yet, only doublets are obtained. It seems to me there should be more multiplicity in the methylene signals.

J. Prud'homme: I think here the same situation arises as with hydrogenated 1,4-polydimethylbutadiene. It would appear that when two methylene units separate two chiral carbons, each methylene unit shows little sensitivity to the meso and racemic configurations of the two adjacent chiral carbons. Carman et al. have reported spectra measured on alkanes which show that when two tertiary chiral carbons are separated by two methylene units, the difference in the chemical shifts of the methylene units in the meso and racemic configurations is close to 0.3 ppm. It was not possible to observe this kind of resolution in the spectra of the present polymers. Only a broadening effect occurred.

RECEIVED March 13, 1979.

# Carbon-13 NMR Studies on the Cationic Polymerization of Cyclic Ethers

G. PRUCKMAYR and T. K. WU

E. I. du Pont de Nemours & Co., Inc., Experimental Station, Wilmington, DE 19898

The cationic ring-opening polymerization of cyclic ethers has been the subject of many recent investigations (1,2,3,4). Nuclear magnetic resonance (NMR) methods, particularly carbon-13 techniques, have been found most useful in studying the mechanism of these polymerizations (5). In the present review we would like to report some of our recent work in this field.

The first part of this report will illustrate how [13]C-NMR has been utilized in the elucidation of the polymerization mechanisms of cyclic ethers. In the second part, quantitative applications of [13]C-NMR for determinations of thermodynamic and kinetic constants will be discussed. The last section deals with possible applications of quantitative [13]C-NMR analysis in copolymerization of cyclic ethers.

## EXPERIMENTAL

Tetrahydrofuran (THF) and oxepane (OXP) were distilled from $CaH_2$ prior to use. All other reagents and solvents are commercially available in reagent grade purity and were used without further purification.

The proton noise-decoupled [13]C-NMR spectra were obtained on a Bruker WH-90 Fourier transform spectrometer operating at 22.63 MHz. The other spectrometer systems used were a Bruker Model HFX-90 and a Varian XL-100. Tetramethylsilane (TMS) was used as internal reference, and all chemical shifts are reported downfield from TMS. Field-frequency stabilization was maintained by deuterium lock on external or internal perdeuterated nitromethane. Quantitative spectral intensities were obtained by gated decoupling and a pulse delay of 10 seconds. Accumulation of 1000 pulses with phase alternating pulse sequence was generally used. For "relative" spectral intensities no pulse delay was used, and accumulation of 200 pulses was found to give adequate signal-to-noise ratios for quantitative data collection.

A calibration curve was obtained from [13]C-NMR spectra of a series of polytetramethylene ether (PTME)-THF/$CH_3NO_2$ solutions at

different concentrations and temperatures.  The PTME was obtained
by polymerization of THF with $Me_3O^+BF_4^-$ in $CH_3NO_2$ (molar ratio
$THF:Me_3OBF_4:CH_3NO_2$ = 1.36:0.1:1.07) under conditions similar to
the subsequent kinetic study.  The reaction mixture was quenched
with MeONa/MeOH, and the polymer isolated by removal of unreacted
monomer and solvent under vacuum, and extraction of the residue
with ether.  After isolation, the resulting dimethoxypolytetra-
methylene ether MeO—$[CH_2CH_2CH_2CH_2O]_n$—Me (Mn $\approx$ 600) was used
directly in the calibration mixtures.

RESULTS AND DISCUSSION

I.  Polymerization of Cyclic Ethers

    General Mechanism and Spectra.  The cationic ring-opening
polymerization of cyclic ethers has long been known to involve
oxonium ions (6).  For THF it is well recognized that under
certain conditions all the reactions are reversible and that
limiting conversions are reached at given temperatures.  The
polymerization of THF has therefore been frequently characterized
as a "living" polymerization (7).

    In the initial step of the polymerization, a cyclic oxonium
ion is formed by transfer of an alkyl group from the initiator to
the cyclic ether.  Propagation occurs by $SN_2$ attack of a monomer
molecule at a ring α-methylene position of the cyclic tertiary
oxonium ion, followed by opening of the oxonium ring and formation of a new cyclic oxonium ion.

    The initiator may be a Lewis acid, an oxonium salt or pre-
cursor (8), or an ester of a strong acid (9).  The anion A⁻ in
the formula scheme below may designate, e.g., tetrafluoroborate
($BF_4^-$), fluorosulfonate ($FSO_3^-$), trifluoromethyl sulfonate
($CF_3SO_3^-$), etc.

Kinetic study of this reaction usually requires sampling the polymerizing mixture and analyzing for the concentrations of the various reaction species at different polymerization times. Vofsi and Tobolsky in 1965 reported the use of radioactively tagged initiator (10), while Saegusa and coworkers in 1968 developed a "phenoxy end-capping" method in which the oxonium ion is trapped with sodium phenoxide and the derived phenyl ether at the polymer chain end quantitatively determined by UV spectrophotometry (11).

We have been investigating similar model polymerizations. In 1973 we reported the use of $^1$H-NMR spectroscopy for the identification of the various species in such a polymerization (12). We found this method to be extremely useful for kinetic in-situ study of polymerizations without disturbing the system. Subsequently we applied $^{19}$F-NMR to follow the polymerization initiated with catalysts containing fluorine atoms (13). At the same time the superior resolution of $^{13}$C-NMR was exploited to investigate the various proposed equilibria in the polymerization of cyclic ethers (5).

Figure 1 shows the proton noise-decoupled $^{13}$C-NMR spectrum of a polytetrahydrofuran (polytetramethylene ether glycol, PTMEG) dissolved in THF. In this spectrum the carbons numbered 1, 2 and 3 which are $\alpha$ to the oxygen appear at lower field than the $\beta$-carbons labeled as 4, 5 and 6. The carbon atoms in the polymer are clearly resolved from the corresponding carbons of the THF monomer. The fact that carbons 3 and 4 near the hydroxyl endgroups can be easily identified shows the excellent resolution of this technique.

Polymerization Equilibria. As mentioned earlier, esters of strong acids, e.g. trifluoromethane sulfonic acid ("triflates"), are excellent initiators for the polymerization of THF. With such initiators, however, a complication arises. In addition to the normal propagation $\rightleftarrows$ depropagation equilibria of oxonium ions, Smith and Hubin postulated that the macroion (I) may also convert into a corresponding nonpolar macroester (E) by attack of the anion (14).

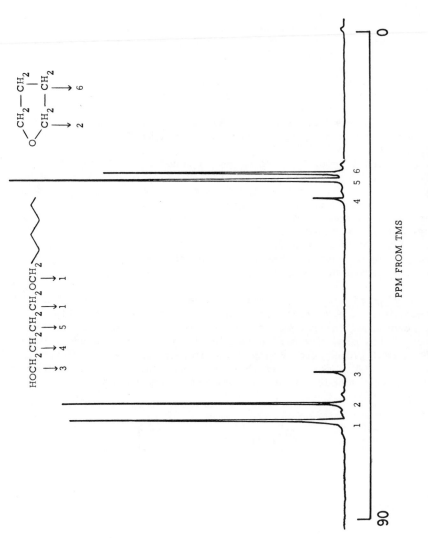

*Figure 1.   C-13 NMR spectrum of PTMEG in THF (1:1)*

Subsequently, Penczek further expanded this concept and concluded that the extent of macroion or macroester formation was dependent on the polarity of the polymerization medium (15). Initial efforts to substantiate this theory using proton NMR did not lead to unambiguous spectral assignments (11). By recognizing the large chemical shift difference between fluorine in an anion and in a neutral species, Saegusa's group and we independently obtained $^{19}F$-NMR data to support this mechanism (13,16). However, the $^{19}F$ technique is limited to examination of fluorine-containing initiators, and we decided to use $^{13}C$-NMR to shed further light into the nature of this problem.

Figure 2 shows the complete $^{13}C$ spectra of two THF/methyltriflate polymerization mixtures, one in a nonpolar solvent ($CCl_4$, Figure 2,B) and the other one in a strongly polar solvent ($CH_3NO_2$, Figure 2,A). The chemical shifts of the α- and β-methylene carbon peaks of the polymer and those of the monomer closely correspond to those shown in the spectrum of polytetrahydrofuran (Figure 1). It is noteworthy that at the low field side of this polymer peak two strong signals appear at 87.5 ppm in $CH_3NO_2$ solution, while a single resonance peak is observed at 79.1 ppm in $CCl_4$ solution. We assigned the two low-field signals in $CH_3NO_2$ to the endo- and exo-cyclic α-methylene carbons at the oxonium ion, respectively, and the single low-field signal in $CCl_4$ to the α-methylene carbon of the corresponding macrofluorosulfate (5). (Some of the spectral assignments were confirmed by off-resonance decoupling.) The $^{13}C$-NMR spectra therefore supported the macroion-macroester equilibration proposed by Smith and Hubin (14), and Penczek (15).

Further detail can be seen in Figure 3, which is a horizontal expansion of the oxonium region from about 85 ppm to 95 ppm, illustrating the spectra of reaction mixtures of methyltriflate with 5-, 6-, and 7-membered ring compounds. The α-methylene-carbon peaks of the methyl oxonium ions of the 5- and 7-membered cyclic ethers are found at the low field side of the spectra. A consistent ring-size effect is evident resulting in a downfield shift of about 1.7 ppm per ring expansion by one $CH_2$ unit. In compounds which undergo ring-opening polymerization the chemical shift of the open chain or "exo-cyclic" methylene carbons of the polymeric oxonium ions is different from the chemical shift of the ring, or "endo-cyclic" methylene carbons. Tetrahydropyran (THP), the strainless 6-membered ring, forms a tertiary oxonium ion, but does not subsequently ring open.

Based on $^1H$-, $^{19}F$-, and $^{13}C$-NMR results we have schematically represented the equilibrium polymerization of THF with esters of trifluoromethane sulfonic acid as shown in Scheme I. Initiation occurs when the alkyl (R) group of the ester is transferred to THF to form an oxonium ion. In polar media, the oxygen atom of another THF molecule will add to the α-methylene position of the oxonium ion leading to ring-opening propagation. Because the charged species are stabilized in polar medium, the

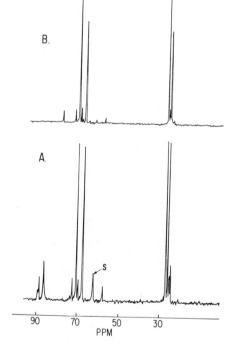

Figure 2. C-13 NMR spectra (22.63 MHz) of the polymerization mixture of THF–CH₃OSO₂F (6:1): (A) 64% in CH₃NO₂, after a polymerization time of 20 min. (S indicates the solvent peak); (B) 64% in CCl₄, after a polymerization time of 20 min.

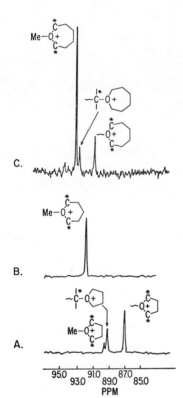

*Figure 3. C-13 NMR spectra (22.63 MHz) of the oxonium ion region: (A) THF–CH₃OSO₂CF₃ (6:1) in CH₃NO₂ (64%), after 15 min; (B) THP–CH₃-OSO₂CF₃ (6:2) in CH₃NO₂ (67%), after 60 min; (C) OXP–CH₃OSO₂CF₃ (2:6) in CH₃NO₂ (67%), after 30 min.*

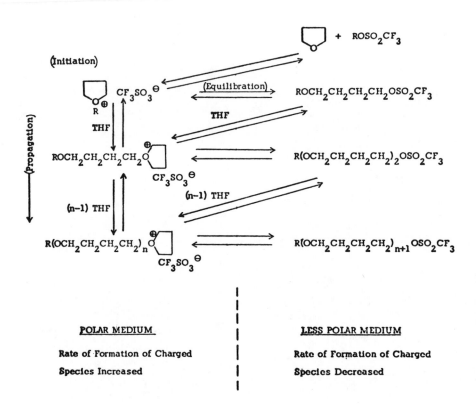

*Reaction Scheme I.  Polymerization of THF with esters of fluorosulfonic acids*

equilibrium is shifted to favor the macroion. Therefore, the polymerization proceeds largely by following the steps from top to bottom shown on the left side of the outline.

In nonpolar media, on the other hand, the newly formed oxonium ion will either quickly convert to the corresponding soluble ester, or it will precipitate, since monomeric or short-chain oligomeric oxonium salts have low solubility in such media. The soluble ester is structurally similar to the initiator and may add another THF molecule. The resulting oxonium ion will again revert to the ester or precipitate. In fact, precipitates are generally observed durina the early stages of polymerization in media of low polarity. They have been isolated and characterized as monomeric or short chain oligomeric oxonium salts (17).

As the polymer chains increase in length (at longer polymerization times or very low initiator concentrations), they will tend to stabilize the ionic ends in solution. Although the concentration of ionic species under these conditions will still be very low, both types of end-groups may participate in chain propagation (18), since the propagation rate of oxonium ions was found to be much higher than that of the corresponding macroester (19).

Chain Transfer. Dreyfuss and Dreyfuss discussed the possibility of chain transfer during cationic polymerization of cyclic ethers (20). This can occur when the cyclic oxonium ion is attacked by an oxygen of a polymer molecule rather than by monomer. The oxonium ion formed in this case is a branched site, an open-chain tertiary oxonium ion, which has been called a "dormant" ion because of its lack of ring strain (21).

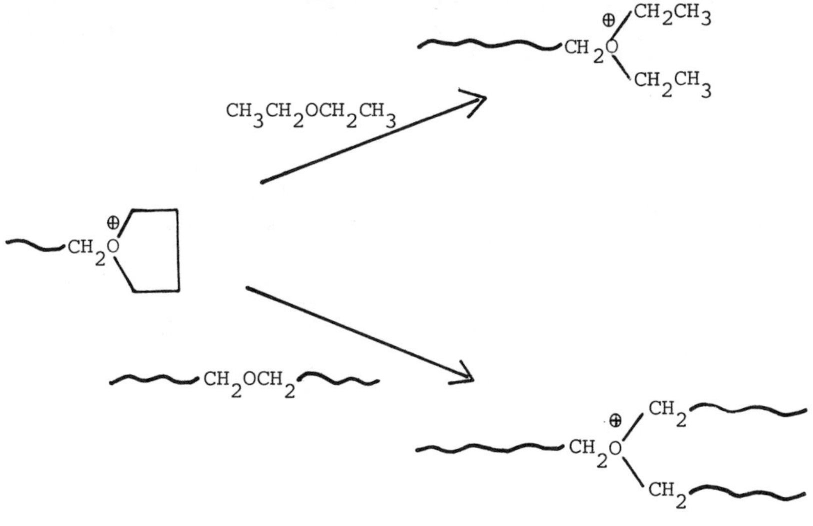

"DORMANT"

Dreyfuss and Dreyfuss reasoned that a similar chain transfer should also occur with added small acyclic ethers.  Indeed in the presence of diethyl ether they found that the ultimate conversion of THF to polymer was not affected but that the intrinsic viscosity of the polymer decreased with time (20).

Investigation of this chain transfer reaction is greatly facilitated by using [13]C-NMR.  In Figure 4 the low field region of a polymerization mixture of THF/MeOSO$_2$F/diethyl ether is shown. We observe the $\alpha$-methylene carbons of the methyl tetrahydrofuranium ion, the $\alpha$-carbons of the two types of propagating chain heads, the macroion and the macroester (17).  The observation of the $\alpha$-methylene carbon resonances of the acyclic tertiary oxonium ion provides a direct proof of chain transfer reaction in THF polymerization.

Formation of Cyclic Oligomers.  Chain transfer reactions occur by intermolecular attack of oxygen from another polyether chain on the $\alpha$-methylene carbons of the oxonium ion.  In an intramolecular attack a distant oxygen of the growing polymer chain itself attacks the $\alpha$-methylene position of its oxonium center. In this case a macrocyclic oxonium ion is formed.  Subsequent exocyclic attack by a monomer molecule will yield a macrocyclic compound containing more than one monomer units (Scheme II).

We first confirmed the formation of these macrocycles in the polymerization of THF by using coupled gas chromatography/mass spectrometry (22).  Macrocyclic ethers containing up to 8 THF units could be separated and identified by this method (23).  The two predominant macrocyclic species found in THF polymerization mixtures are a cyclic tetramer and a cyclic pentamer.  In analogy to the "crown ether" nomenclature, we proposed the name 20-crown-4 for the cyclic tetramer and 25-crown-5 for the cyclic pentamer (22).

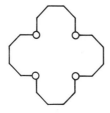

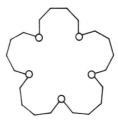

20-CROWN-4                          25-CROWN-5

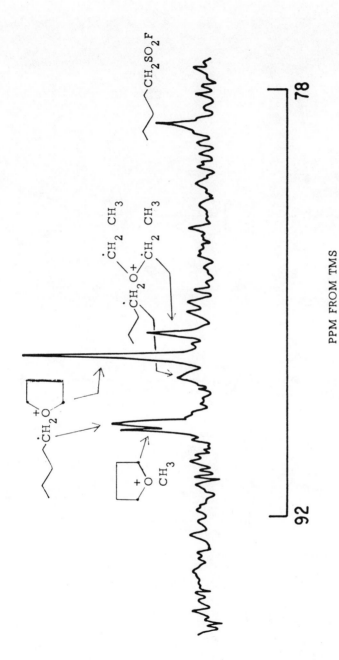

*Figure 4.  Partial C-13 NMR spectrum of a THF–CH₃OSO₂F (2.5:1) polymerization mixture in CH₃NO₂ (45.5%) after addition of diethyl ether (2.5)*

PPM FROM TMS

INTERMOLECULAR ATTACK:

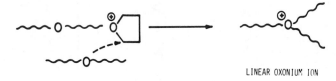

LINEAR OXONIUM ION

INTRAMOLECULAR ATTACK:

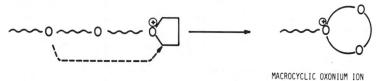

MACROCYCLIC OXONIUM ION

*Reaction Scheme II.*

Identification of these macrocycles was also facilitated by examining their [13]C-NMR spectra. Figure 5 shows a spectrum of the GC fraction of 20-crown-4. Due to the symmetry of this molecule there are only two distinguishable carbons: They are those α and β to the oxygen atoms at 70.4 ppm and 26.5 ppm, respectively. (The triplet at 77 ppm is due to the solvent, $CDCl_3$.)

Comparison of the chemical shift data (Table 1) reveals that the peak positions of α and β carbons of 20-crown-4 are quite different from the corresponding carbons of THF or the polymeric PTME. Small but distinct chemical shift differences were also found for macrocyclic oligomers of other ring sizes.

## II.  Quantitative Applications of [13]C-NMR

In view of the excellent resolution of [13]C spectra it would be of interest to use these data for quantitative correlations. However, quantitative analysis by proton noise-decoupled Fourier transform [13]C-NMR is complicated by the fact that different carbon nuclei may have different spin relaxation times and nuclear Overhauser enhancement (NOE) factors. Therefore, the observed peak areas in the spectra are not necessarily proportional to the number of carbon atoms involved.

Schaefer and Natusch have shown that for many synthetic high polymers in solution the NOE factors and relaxation times of carbon atoms in or near the main chains are similar (24). In such cases the relative peak areas in the spectra obtained by the noise-decoupled and fast pulsing technique can be used as a good approximation for quantitative microstructure analysis. However for our investigation of the polymerization of cyclic ethers we are frequently interested in the quantitative measurements of monomers and oligomers as well as the concentrations of the continuously growing polymeric species. Therefore, the assumption of Schaefer and Natusch is not applicable.

The standard method of obtaining quantitative spectra involves the use of gated decoupling and long pulse delay, both of which require very long data collection times. Figure 6 depicts the partial [13]C spectrum of the α-carbon region of an equilibrated polymerization mixture of $THF/Me_3OBF_4$ in $CD_3NO_2$. Gated decoupling and a long pulse delay time of 10 seconds were employed to obtain the spectrum. From the monomer and polymer peak areas, the extent of polymerization at equilibrium can be determined. Measurements of chain end and the polymer peaks provide information on number-average degree of polymerization. The data collection time required to obtain this spectrum was almost three hours.

Since we were also interested in obtaining quantitative kinetic data for which the long data collection time technique cannot be used, we devised a second approach using "relative" peak intensities in the spectra obtained by fast pulsing. The two approaches are summarized as follows:

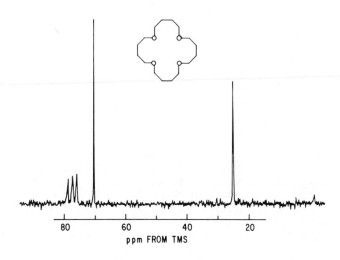

*Figure 5.   C-13 NMR spectrum of the cyclic THF tetramer 20-crown-4 in CDCl₃*

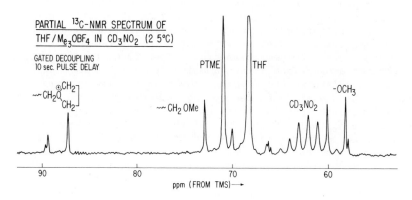

*Figure 6.   Polymerization mixture THF–CD₃NO₂–Me₃OBF₄ (mol ratios =
1:0.75:0.08) after equilibration*

Table I.  C-13 NMR Chemical Shift of Tetramethylene Ethers

|  | α-CARBONS | β-CARBONS |
|---|---|---|
| THF | 68.2 | 26.2 |
| 20-CROWN-4 | 70.4 | 26.5 |
| $\sim\sim CH_2OCH_2(CH_2)_2CH_2OCH_2\sim\sim$ | 71.1 | 27.4 |

**Table II.  Quantitative C-13 NMR Spectroscopy**

A.  "Absolute" Signal Intensities (Long Data
     Collection Times)
          Gated Decoupling to Suppress NOE
          Pulse Delay (>10 seconds)

B.  "Relative" Signal Intensities (Short Data
     Collection Times)
          Internal Standard
               Assumptions:
                    No Change in Relaxation Time
                    No Change in NOE
                    Viscosity Effects Negligible


In the first and obvious approach, "absolute" signal intensities are measured.  Since very long data collection times are
required, this method is only useful in studying equilibrated,
i.e. nonchanging, systems.

In the second approach, "relative" signal intensities are
compared, and data collection times of the order of 5 to 8 minutes
per scan were found to be sufficient.  In this approach, an
internal standard peak, such as a solvent peak (e.g. $CH_3NO_2$), is
used as the reference and compared with the peak intensity of a
carbon of interest, e.g. of monomer.  The underlying assumptions
are that the relaxation time and NOE ratios of the internal standard and the carbon of interest remain unchanged during the
course of polymerization, and that viscosity effects are negligi-
ble.  Since we are dealing with relatively low conversions and
low molecular weight polymers in solution, this assumption is
not unreasonable.

In order to verify the validity of these assumptions we prepared several calibration samples containing different ratios
of THF to $CH_3NO_2$.  Different amounts of polymer were added to
these samples to simulate the viscous properties of the polymerization mixture.  We found that the peak intensity ratio of THF
to $CH_3NO_2$ obtained by the fast pulsing technique can indeed be
linearly correlated with the corresponding weight ratios of these
two compounds.  Moreover, change of temperature from 0 to 35°
introduced no appreciable deviations.  The calibration curve is
shown in Figure 7.  The composition range of interest in our polymerization study is indicated by the bracket.

Thermodynamic Data.  For an equilibrium polymerization, the
equilibrium constant $K_e$ is equal to the ratio of rate of

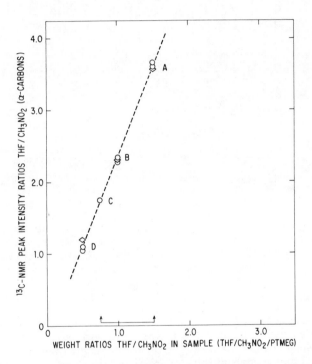

*Figure 7.    C-13 NMR calibration curve: (A), no PTME; (B), 20% PTME; (C), 30% PTME; (D), 40% PTME; (◇), 0°C; (○), 35°C.*

propagation, $k_p$, to that of depropagation, $k_{-p}$. This constant is also related to monomer and polymer concentrations by the law of mass action. To a good approximation $K_e$ is equal to $1/[M]_e$, where $[M]_e$ is the monomer concentration at equilibrium. From the free energy relationship one may rearrange the terms and make appropriate substitution to obtain the expression as shown (Eq. 1). Measurements of $[M]_e$ at different polymerization temperatures should yield the enthalpy and entropy of polymerization.

$$K_e = \frac{k_p}{k_{-p}} = \frac{1}{[M]_e}$$

$$\Delta F_p = -RT \ln K_e$$

$$= \Delta H_p - T\Delta S_p$$

$$\boxed{\ln [M]_e = \frac{1}{T}\frac{\Delta H_p}{R} - \frac{\Delta S_p}{R}} \qquad (1)$$

In Figure 8 the log of $[M]_e$ is plotted against the reciprocal of polymerization temperature. Three types of NMR data are shown. The filled circles are from the fast pulsing technique, the open circle from the gated and delayed pulsing technique and the open squares from proton NMR. From the slope and intercept of the least-square fitted line the enthalpy and entropy of polymerization were obtained, respectively.

The thermodynamic constants of THF polymerization have been investigated by a number of authors. A variety of experimental techniques have been utilized including determinations of conversion to polymer, combustion, heat capacities and vapor pressure. Comparison of our results with some previously published data shows that our results are within the range of the values reported (Table 3).

Kinetic Data. Let us now consider the kinetics of a reversible cyclic ether polymerization. For such a polymerization in progress, the kinetic expression is

$$-\frac{d[M]}{dt} = k_p [P^*][M] - k_{-p}[P^*] \qquad (2)$$

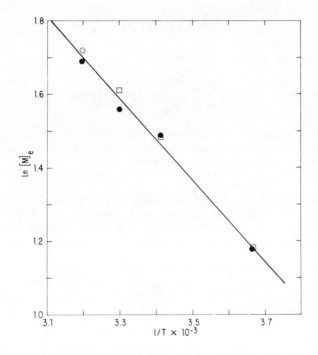

*Figure 8. Determination of thermodynamic constants of THF polymerization (plot of Equation 1): (●), C-13 NMR (decoupled and fast pulsing); (○), C-13 NMR (gated and delayed pulsing); (□), $^1H$ NMR. $\Delta H_P = -2.2$ kcal mol$^{-1}$; $\Delta S_P = -10.4$ cal deg$^{-1}$ mol$^{-1}$.*

where $\boxed{M}$ and $\boxed{P^*}$ are the molar concentrations of the monomer and growing polymer respectively, and $k_p$ and $k_{-p}$ are the rate constants defined earlier.

At equilibrium $d\boxed{M}/dt = 0$ and

$$k_p[M]_e = k_{-p} \tag{3}$$

By proper substitution, Eq. (2) is simplified to

$$-\frac{d[M]}{dt} = k_p[P^*]\left\{[M]-[M]_e\right\} \tag{4}$$

Integration of Eq. (4) leads to

$$\ln\frac{[M]_{t_1}-[M]_e}{[M]_{t_2}-[M]_e} = k_p\int_{t_1}^{t_2}[P^*]\,dt \tag{5}$$

If the instantaneous monomer concentrations $\boxed{M}_{t1}$ and $\boxed{M}_{t2}$ can be continuously monitored during polymerization, and $\boxed{P^*}$ is also known, $k_p$ can then be calculated with Eq. (5). This approach was used by Saegusa and others to study the polymerization of THF (25,26).

However, this relationship is not applicable when data from $^{13}$C spectra are used. $^{13}$C spectra are obtained via Fourier transform computation of data accumulated over a definite time interval, and instantaneous concentration measurements are not possible.

We have therefore modified the kinetic expression to handle the $^{13}$C-NMR data. For the special case where the concentration of active chain ends $\boxed{P^*}$ is constant, the derived kinetic expression is reduced to a very simple form:

$$\boxed{\ln\left([\bar{M}]_t-[M]_e\right) \cong -k_p[P^*]t + CONSTANT} \tag{6}$$

$\boxed{\bar{M}}_t$ represents the monomer concentration at time t, as obtained from Fourier transform $^{13}$C-NMR data. The rate constant of propagation $k_p$ can now be determined by measuring $\boxed{\bar{M}}_t$ as a function of polymerization time, t. (For the derivation of this expression, see the Appendix.)

In a kinetic study, we carried out a polymerization reaction of THF in $CH_3NO_2$ with $(CH_3)_3\overset{+}{O}BF_4^-$ at 40°C. to equilibrium and then quickly chilled the reaction mixture to 0°C. to follow further polymerization at this temperature. The kinetic data obtained are shown in Table 4. A $^{13}$C scan was obtained once every 8 minutes which was found to be the optimal spectral accumulation time.

The data of Table 4 were now plotted in terms of $\ln(\boxed{\bar{M}}_t - \boxed{M}_e)$ versus polymerization time t, counting from the reference

Table III.   Summary of Published Data of Enthalpy and
Entropy of Polymerization of THF

| $-\Delta H_p$ KCal / mol | $-\Delta S_p$ Cal / deg / mol | METHOD | REFERENCE |
|---|---|---|---|
| 2.2 | 10.4 | EQUILIBRIUM ($^{13}$C-NMR) | THIS WORK |
| 4.6 | 17.7 | EQUILIBRIUM (CONV. to POL.) | IVIN et al (1958) |
| 9.1 | — | COMBUSTION | CASS (1958) |
| 4.3 | 17.0 | EQUILIBRIUM (CONV. to POL.) | SIMS (1964) |
| 3.3 | 10.7 | EQUILIBRIUM WITH SOLVENT CORRECTION | IVIN et al (1965) |
| — | 14.8 | HEAT CAPACITIES | GLEGG et al (1968) |
| 1.8 | 3.9 | EQUILIBRIUM VAPOR PRESSURE | BUSFIELD et al (1972) |

Table IV.   Polymerization of Tetrahydrofuran[1] in Nitromethane at 0°C

| t (min) | $[\bar{M}]_t^{2}$ | $\ln\left([\bar{M}]_t - [\bar{M}]_e\right)$ |
|---|---|---|
| 4 | 2.57 | 0.25 |
| 12 | 2.09 | $-0.21$ |
| 20 | 1.99 | $-0.34$ |
| 28 | 1.74 | $-0.77$ |
| 42 | 1.50 | $-1.51$ |
| 58 | 1.36 | $-2.53$ |
| EQUILIBRIUM | 1.28 | — |

[1] THF: 7.38 mol $l^{-1}$, $(CH_3)_3OBF_4$: 0.59 mol $l^{-1}$
[2] INTEGRATED MONOMER CONCENTRATION IN RELATIVE UNITS

time $t_o$ (Figure 9). These data were then least-square fitted to
a straight line. The slope of this line is equal to the negative
value of product $\boxed{P^*}$ and $k_p$. The propagation constant $k_p$ of
THF polymerization in $CH_3NO_2$ at 0°C. was found to be 1.5 x
$10^{-3}$l.mol$^{-1}$sec$^{-1}$. This is in good agreement with the propagation
constant of a similar polymerization mixture at this temperature
calculated from $^{19}$F-NMR data (18).

## III.  Application to Copolymerizations

$^{13}$C-NMR kinetic analysis would appear to be most useful for
studying polymerization systems which cannot be adequately characterized by proton or fluorine NMR methods.  Examples of such
systems are e.g. copolymerizations of cyclic ethers, and in the
last part of this review we would like to discuss briefly some
preliminary results on THF copolymerizations.

Figure 10 presents a summary of the α carbon chemical shifts
of oxonium ions and esters of some of the compounds discussed
earlier.  The carbon atoms α to an oxonium center cover a range
of about 25 ppm.  The peaks due to all the different oxonium ions
and esters can be clearly distinguished, and $^{13}$C-NMR therefore
appeared to be an excellent technique for studying such copolymerizations.

As an example of a cyclic ether copolymerization, we will
briefly discuss the polymerization of THF with OXP initiated with
methyltriflate.  The homopolymerizations of both cyclic monomers
follow a similar mechanism, and both were found to proceed via
macrooxonium ion and/or the macroester mechanism depending on the
polarity of the polymerization medium.  There should then be 8
possible end-groups, i.e. two types of methoxy tails having a
penultimate THF or OXP unit, respectively, two covalent macro-
esters, and four different oxonium ion propagating chain heads:
two from a THF oxonium center attached to penultimate THF or OXP
units, and two from an OXP oxonium center attached to THF and OXP
penultimate units (Scheme III).

Figure 11 shows the α carbon resonance region of such a THF/
OXP copolymerization in $CH_3NO_2$.  At about 55 ppm we observe the
peak due to the methoxy methyl carbons of the chain ends, and
further downfield a solvent peak and then the methylene carbons of
the unreacted monomers, THF and OXP.  There are two peaks attributable to the polymeric methylene carbons.  The higher field one
is due to THF and the other one to OXP.  Similarly, two peaks are
observed for the methylene carbons attached to the methoxy chain
ends.  The fact that the intensities of these two peaks are similar indicates that both THF and OXP participate in the initiation
step.

In the macroester region, we have a small signal due to OXP-
ester only, no THF macroester was observed under these conditions.
On the other hand, in the oxonium region there are resonances due
to the THF-macroions only but not to OXP-macroions.  Therefore,
in this particular polymerization system we have identified 4 out

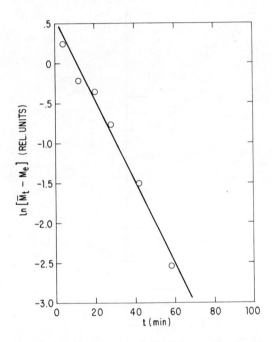

*Figure 9.   Determination of kinetic constants (plot of Equation 6). Polymerization of THF in $CH_3NO_2$ at $0°C$: $k_P = 1.5 \times 10^{-3}$ L mol$^{-1}$ sec$^{-1}$.*

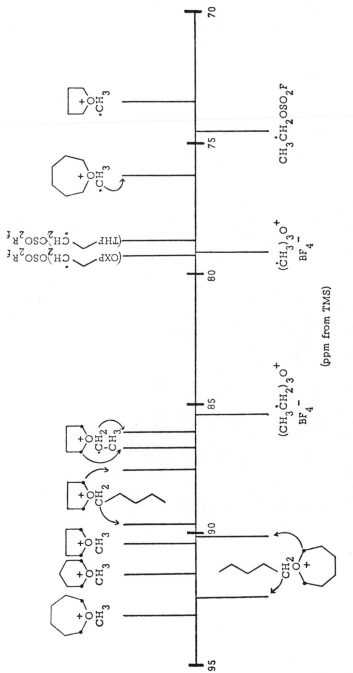

Figure 10.    C-13 NMR chemical shift assignments of some oxonium ions and fluorosulfonate esters

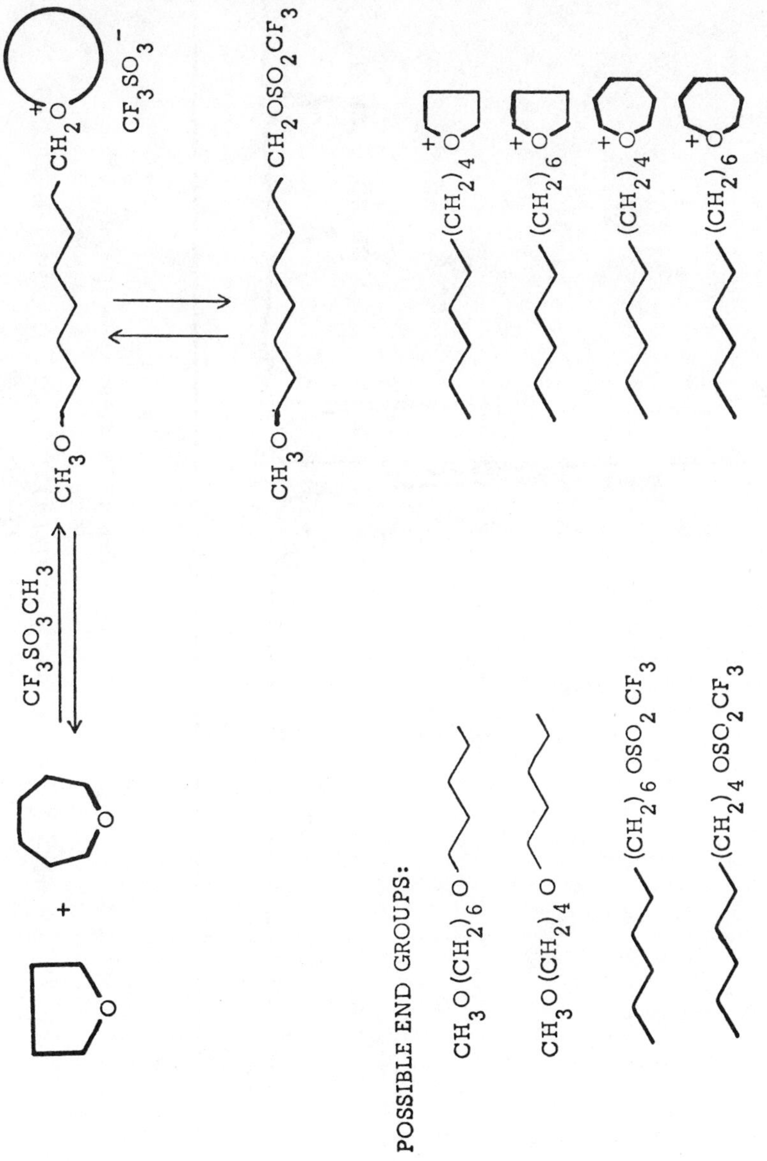
CF3SO3CH3
CH3O
CH2
CF3SO3
CH3O
CH2OSO2CF3
POSSIBLE END GROUPS:
CH3O(CH2)6O
CH3O(CH2)4O
(CH2)6 OSO2 CF3
(CH2)4 OSO2 CF3
(CH2)4
(CH2)6
(CH2)4
(CH2)6

*Reaction Scheme III. Copolymerization*

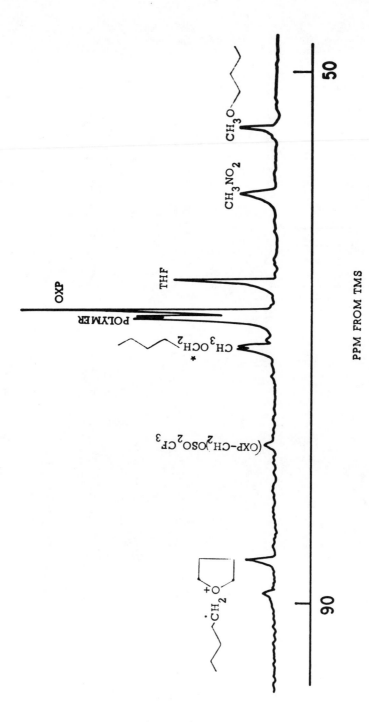

Figure 11.   Partial C-13 NMR spectrum of a THF–OXP–$CF_3SO_3CH_3$ (1.4:2.8:1) polymerization mixture in $CH_3NO_2$ (28%) after 30 min at 25°C

of the 8 possible chain ends. The results of a quantitative [13]C
analysis of a similar THF/OXP copolymerization system are summarized in Table 5 in terms of total % conversion to polymer,
kinetic degree of copolymerization, copolymer composition and
feed composition. The kinetic degree of copolymerization of 6.2
is, within experimental error, the same as the theoretically
calculated value. Therefore, our results indicate that each
initiator molecule initiates one copolymer chain as in homopolymerizations. Furthermore the copolymer composition was found to
be very similar to feed composition. This suggests that cationic
copolymerization of the two cyclic ethers may be statistically
random.

**Table V.    Copolymerization of THF–OXP at 20°C**

Moles of $THF/OXP/CD_3NO_2/CH_3SO_3CF_3$ = 1.04/.75/1.56/.14

| | |
|---|---|
| Total Conversion, % | 44 |
| Kinetic Degree of Copolymerization | 6.2 |
| Copolymer Composition | |
|     Mol % THF | 57 |
|     Mol % OXP | 43 |
| Feed Composition | |
|     Mol % THF | 58 |
|     Mol % OXP | 42 |

We are currently extending this approach to investigate the
copolymerization of THF with other cyclic ethers.

CONCLUSION

NMR methods are ideally suited to study polymerization
reactions of cyclic ethers in situ, without disturbing the polymerization equilibria. [13]C-NMR methods generally afford much
more detailed information than either [1]H- or [19]F-NMR methods, but
quantitative evaluation of data is not as straightforward, due
to the necessity for Fourier transform and noise decoupling techniques. A [13]C-NMR method based on relative signal intensities
has now been developed for obtaining quantitative information of
homopolymerization and copolymerization systems, which may otherwise not be easily accessible.

ACKNOWLEDGMENT

We would like to thank Dr. W. W. Yau for his assistance in
modifying the kinetic theory, Dr. J. J. Chang for technical assistance, and Dr. G. E. Heinsohn for a critical review of the
manuscript.

<u>APPENDIX</u>

<u>Kinetic Analysis of Cyclic Ether Polymerization
by Fourier Transform NMR</u>

The kinetic expression generally applicable to reversible equilibrium polymerizations is:

$$\ln \frac{[M]_{t_1} - [M]_e}{[M]_{t_2} - [M]_e} = k_p \int_{t_1}^{t_2} [P^*]\, dt \tag{1}$$

$[M]_t$ is the monomer concentration at time t, $[M]_e$ is the equilibrium monomer concentration, and $[P^*]$ is the concentration of active polymer chain sites. Application of this formula requires instantaneous concentration measurements at times $t_1$ and $t_2$. This relationship is not applicable when the data are accumulated over a definite time interval, such as by multiple pulsing in [13]C-NMR. In order to handle this type of data, we have modified this expression, introducing an arbitrary reference time, $t_o$. $[M]$ and $[M]_o$ are the monomer concentrations at times t and $t_o$.

$$-\ln \frac{[M] - [M]_e}{[M]_o - [M]_e} = k_p \int_{\tau = t_o}^{t} [P^*]\, d\tau = k_p\, P(t) \tag{2}$$

P(t) is used to designate the integral of $[P^*]$. Eq. (2) is then converted into an exponential form and integrated through $\Delta t$, the time interval required for spectral accumulation.

$$\int_{\tau = t - \frac{1}{2}\Delta t}^{t + \frac{1}{2}\Delta t} [M]\, d\tau - [M]_e\, \Delta t$$

$$= \left([M]_o - [M]_e\right) \int_{\tau = t - \frac{1}{2}\Delta t}^{t + \frac{1}{2}\Delta t} e^{-k_p P(\tau)}\, d\tau \tag{3}$$

Rearrangement of terms of Eq. (3) leads to

$$\frac{[\overline{M}]_t - [M]_e}{[M]_o - [M]_e} = \frac{1}{\Delta t} \int_{\tau = t - \frac{1}{2}\Delta t}^{t + \frac{1}{2}\Delta t} e^{-k_p P(\tau)}\, d\tau \tag{4}$$

The $[\overline{M}]_t$ represents the integral and is the monomer concentration measured from the [13]C spectrum at time t.

Since the integral of the right-hand term of Eq. (4) cannot be analytically integrated, we took the following procedure:

Let the integrand of that integral be X and the integrated X be Y. If each Y is expanded in Taylor's series, we obtain two series of terms in Eq. (5).

$$\int_{\tau \,=\, t-\frac{1}{2}\Delta t}^{t+\frac{1}{2}\Delta t} X d\tau = Y\left(t + \frac{1}{2}\Delta t\right) - Y\left(t - \frac{1}{2}\Delta t\right)$$

$$= Y(t) + \left(\frac{1}{2}\Delta t\right)Y' + \frac{\left(\frac{1}{2}\Delta t\right)^2}{2!} Y'' + \frac{\left(\frac{1}{2}\Delta t\right)^3}{3!} Y''' + \cdots$$

$$- \left[ Y(t) - \left(\frac{1}{2}\Delta t\right)Y' + \frac{\left(\frac{1}{2}\Delta t\right)^2}{2!} Y'' - \frac{\left(\frac{1}{2}\Delta t\right)^3}{3!} Y''' + \cdots \right]$$

$$= (\Delta t)Y' + \frac{(\Delta t)^3}{24} Y''' + \frac{(\Delta t)^5}{16 \times 5!} Y'''' + \cdots$$

$$= (\Delta t)\left[ X + \frac{(\Delta t)^2}{24} X'' + \cdots \right] \tag{5}$$

where Y', Y" and Y'" represent the first, second and third derivatives of Y, respectively. Since every other term of these two series cancels, we collect the remaining terms in Y's and then in X's. The X's and Y's are related by the expressions shown below:

$$X = Y' = e^{-k_p P(t)} \tag{6}$$

$$X' = Y'' = \frac{dX}{dt} = -k_p [P^*] X \tag{7}$$

$$X'' = Y''' = \left( k_p^2 [P^*]^2 - k_p \frac{d[P^*]}{dt} \right) X \tag{8}$$

In Eq. (9) we write the kinetic equation first in terms of X and then in terms of the pertinent kinetic constants.

$$\frac{[\bar{M}]_t - [M]_e}{[M]_0 - [M]_e} = X + \frac{(\Delta t)^2}{24} X'' + \cdots$$

$$= e^{-k_p P(t)} \left\{ 1 + \frac{(\Delta t)^2}{24} \left( k_p^2 [P^*]^2 - k_p \frac{d[P^*]}{dt} \right) \right\} + \cdots \tag{9}$$

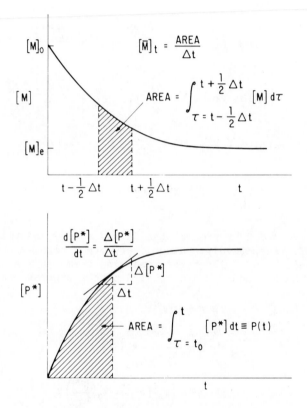

*Figure 12.    Model graphs for a kinetic analysis of an equilibrium polymerization (definition of variables of Equation 11)*

Next, Eq. (9) is converted into the logarithmic form

$$\ln \frac{[M]_0 - [M]_e}{[\bar{M}]_t - [M]_e} = k_p P(t) - \ln \left\{ 1 - \frac{(\Delta t)^2}{24} k_p \frac{d[P^*]}{dt} + \frac{(\Delta t)^2}{24} k_p^2 [P^*]^2 + \bullet \bullet \bullet \right\} \quad (10)$$

The higher order terms in $\Delta t$ of Eq. (10) are very small and can be neglected, and we arrive at the final expression.

$$\ln \frac{[M]_0 - [M]_e}{[\bar{M}]_t - [M]_e} \cong k_p \left[ P(t) + \frac{(\Delta t)^2}{24} \frac{d[P^*]}{dt} \right] - \frac{(\Delta t)^2}{24} k_p^2 [P^*]^2 \quad (11)$$

Examination of Eq. (11) reveals that it is very similar to the well known equation used with instantaneous data collection, but it contains two additional correction terms. These correction terms will vanish if $\Delta t$, the accumulation time, becomes infinitesimally small. Then $[\bar{M}]_t$ becomes the instantaneous monomer concentration.

Figure 12 illustrates the experimental data required for a kinetic analysis. We need $[M]_0$, the monomer concentration at a reference time $t_0$, the equilibrium concentration $[M]_e$, $[\bar{M}]_t$, which is this area divided by $\Delta t$. Also needed are $P(t)$, the integrated area under the $[P^*]$ versus t curve and the slope, $d[P^*]/dt$.

For polymerization systems in which the concentration of active chain sites $[P^*]$ is constant, the kinetic expression derived in Eq. (11) can be further simplified. Since $d[P^*]/dt = 0$, the second term of Eq. (11) vanishes and the third term becomes a constant. By taking the constant $[P^*]$ outside the integral and integration, the kinetic expression is reduced to the simple form shown in Eqs. (12) and (13) and in the discussion section.

$$\ln \frac{[M]_{t_1} - [M]_e}{[\bar{M}]_t - [M]_e} \cong k_p \int_{\tau = t_1}^{t} [P^*] d\tau + \text{CONSTANT}$$

$$\cong k_p [P^*] (t - t_1) + \text{CONSTANT} \quad (12)$$

$$\ln \left( [\bar{M}]_t - [M]_e \right) \cong -k_p [P^*] t + \text{CONSTANT} \quad (13)$$

Eq. (13) shows that $k_p$ can be determined by measuring $[\bar{M}]_t$ as a function of polymerization of time, t.

## Literature Cited

1.  Saegusa, T., Kimura, Y., Fujii, H., Kobayashi, S., Macromolecules, (1973), 6, 657.

2.  Penczek, St., Matyjaszewski, K., J. Polym. Sci., Symposium 56, (1976), 255.

3.  Hoene, R., Reichert, K. W., Makromol. Chem., (1976), 177, 3545.

4.  Pruckmayr, G., Wu, T. K., Macromolecules, (1978), 11, 662.

5.  Pruckmayr, G., Wu, T. K., Macromolecules, (1975), 8, 954.

6.  Meerwein, H., Delfs, D., Morschel, H., Angew. Chem., (1960), 24, 927.

7.  Dreyfuss, P., Dreyfuss, M. P., Advan. Chem. Ser., (1969), 91, 335.

8.  Dreyfuss, P., Dreyfuss, M. P., "Ring-Opening Polymerizations", K. C. Frisch, S. L. Reegen, eds., Marcel Dekker, N.Y., (1969).

9.  Saegusa, T., Kobayashi, S., Polyethers, ACS Symposium 6, (1975), 150.

10. Vofsi, D., Tobolsky, A. V., J. Polym. Sci., (1965), A3, 3261.

11. Kobayashi, S., Danda, H., Saegusa, T., Bull. Chem. Soc. Japan, (1973), 46, 3214.

12. Pruckmayr, G., Wu, T. K., Macromolecules, (1973), 6, 33.

13. Wu, T. K., Pruckmayr, G., Macromolecules, (1975), 8, 77.

14. Smith, S., Hubin, A. J., J. Macromol. Sci., Chem., (1973), 7, 1399.

15. Matyjaszewski, K., Penczek, St., J. Polym. Sci., Chem., (1974), 12, 1905.

16. Kobayashi, S., Danda, H., Saegusa, T., Macro-molecules, (1974), 7, 415.

17. Pruckmayr, G., Wu, T. K., ACS Central Regional
    Meeting, Akron, Ohio, (May 1976).

18. Kobayashi, S., Morikawa, K., Saegusa, T.,
    Macromolecules, (1975), 8, 386.

19. Buyle, A. M., Matyjaszewski, K., Penczek, St.,
    Macromolecules, (1977), 10, 269.

20. Dreyfuss, M. P., Dreyfuss, P., J. Polym. Sci,
    (1966), (A-1) 4, 2179.

21. Saegusa, T., Kobayashi, S., Progress in Polymer
    Sci. (Japan), (1973), 6, 107.

22. McKenna, J. M., Wu, T. K., Pruckmayr, G.,
    Macromolecules, (1977), 10, 877.

23. Pruckmayr, G., Wu, T. K., Macromolecules, (1978),
    11, 265.

24. Schaefer, J., Natusch, D. F. S., Macromolecules,
    (1972), 5, 416.

25. Saegusa, T., Kobayashi, S., J. Polym. Sci.,
    Polymer Symposia, (1976), 56, 241.

26. Dreyfuss, P., Dreyfuss, M. P., "Comprehensive
    Chemical Kinetics", C. H. Bamford, C. H. Tipper,
    eds., 259, Elsevier, 15, (1976).

## Discussion

W. Pasika, Laurentian Univ., Ont.: I would like to refer
back to the fact that the compositional ratio is the same as the
feed ratio. You indicated it was a random process. If the mechanism is an $S_N2$ type mechanism then the compositional ratio will
depend very much on the character of the monomer. The character
of the copolymerization could be other than random. In such a
system the compositional ratio will not necessarily be the same
as the feed ratio.

T. K. Wu: True. This is a preliminary report, and we discussed only one data point from one particular feed ratio. The
result may be fortuitous.

W. Pasika: On the other hand it may well be that these particular two monomers have in fact the same reactivity when it
comes to the particular type of mechanism.

D. J. Worsfold, NRC, Ont.: I was glad to see that you were
able to identify and measure the amount of dormant polymer present
in the THF polymerization. The amount of the dormant chain end

would increase during the course of the polymerization as the amount of polymer produced increases and one would think this would have some effect on the first order kinetics of the reaction. Yet it has always been successful to use first order kinetics to describe the disappearance of the monomer. Is it because the amount of dormant chain end is very small or is it that there is some compensatory effect which still gives the first order kinetics in monomer disappearance?

T. K. Wu: We tried to determine the amount of dormant species in homopolymerizations. We can measure the ratio of "exocyclic" vs. "endocyclic" methylene carbon resonance, and if the ratio is not exactly 1 to 2 we can get an estimate of the concentration of dormant ion.

G. Pruckmayr, Du Pont, Delaware: The concentration of dormant ion is low. We can estimate the concentration of macrocyclic plus linear dormant ions from the intensity ratio of the different α-methylene groups. The total is quite small in homopolymerizations, particularly at short reaction times.

P. Sipos, Du Pont, Ont.: In connection with your copolymerization have you seen a difference in the composition of the polymer and the feed ratio when larger oxy rings are used?

T. K. Wu: The largest ring used so far was oxepane, but we are planning to go in the other direction, using four membered rings. However, in this case complications arise because the strained oxetane ring does not undergo equilibrium polymerization.

RECEIVED March 13, 1979.

# A Comparison of Models and Model Parameters for the Interpretation of Carbon-13 Relaxation in Common Polymers

ALAN ANTHONY JONES, GARY L. ROBINSON, and FREDRIC E. GERR

Jeppson Laboratory, Department of Chemistry, Clark University, Worcester, MA 01610

Carbon-13 spin relaxation in polymers is now a common probe of chain dynamics resulting in a proliferation of polymers studied and models used to relate spin relaxation to polymer motion (1). Presently it appears useful to draw comparisons both among the dynamics of various polymers and among the interpretational models. For these comparisons, it is necessary to center attention on polymers which have been thoroughly investigated experimentally. By this we mean it is desirable to have measurments of several different relaxation parameters including the spin-lattice relaxation time $T_1$, the nuclear Overhauser enhancement NOE, and to a limited extent the spin-spin relaxation time $T_2$. Although carbon-13 NMR is a most useful probe of chain dynamics, it is also important to measure relaxation parameters of other nuclei in the same polymer. Both proton and fluorine nuclei are excellent candidates which add information on motions at other frequencies not available from carbon-13 relaxation alone (2). An approach somewhat similar to observing two types of nuclei is the observation of one nucleus at several magnetic field strengths since this also provides information on dynamics in other frequency domains. Lastly it is informative to vary such non-spectroscopic parameters as polymer molecular weight, concentration, temperature, and solvent. If a model can consistently interpret a large amount of spin relaxation data under a variety of experimental conditions, then it is worthwhile to consider in some detail the significance of the model parameters.

## Survey of Interpretational Models

Relaxation parameters are usually written in terms of transition probabilities, W, and spectral densities, J. For carbon-13 nuclei under condition of proton decoupling (3)

$$1/T_1 = W_0 + 2W_{1I} + W_2$$

0-8412-0505-1/79/47-103-271$05.00/0

$$(\text{NOE}) - 1 = \gamma_S(W_2 - W_0)/\gamma_I(W_0 + 2W_{1I} + W_2)$$

$$= T_1\gamma_S(W_2 - W_0)/\gamma_I$$

$$W_0 = \gamma_I{}^2\gamma_S{}^2\hbar^2 J_0(\omega_0)/20r^6$$

$$W_{1I} = 3\gamma_I{}^2\gamma_S{}^2\hbar^2 J_1(\omega_I)/40r^6 \qquad\qquad [1]$$

$$W_2 = 3\gamma_I{}^2\gamma_S{}^2\hbar^2 J_2(\omega_2)/10r^6$$

$$\omega_0 = \omega_S - \omega_I; \; \omega_2 = \omega_S + \omega_I.$$

The carbon nuclei are to be identified with I, the proton nuclei with S, and the carbon-proton internuclear distance with r. The spectral density is the Fourier transform of a correlation function which is usually based on a probabilistic description of the motion modulating the dipole-dipole interactions. The spin-spin relaxation time, $T_2$, is usually written directly as a function of spectral densities (3).

$$1/T_2 = (1/40)\hbar^2\gamma_S{}^2\gamma_I{}^2(J_0(\omega_0) +$$

$$3J_1(\omega_I) + 6J_2(\omega_S + \omega_I) + \qquad\qquad [1a]$$

$$4J_0(0) + 6J_1(\omega_S)).$$

The simplest motional description is isotropic tumbling characterized by a single exponential correlation time (4). This model has been successfully employed to interpret carbon-13 relaxation in a few cases, notably the methylene carbons in polyisobutylene among the well studied systems (5). However, this model is unable to account for relaxation in many macromolecular systems, for instance polystyrene (6) and poly(phenylene oxide)(7, 8). In the latter case, the estimate of the correlation time varies by an order of magnitude between an interpretation based on $T_1$ and an interpretation based on the NOE (7).

The failure of this model led to the application of motional descriptions involving several correlation times. The simplest of these, a two correlation time model, was developed by Woessner (9) and suggested for macromolecular systems by Allerhand, Dodrell, and Glushko (3). The model considers two motions modulating the dipole-dipole interaction: anisotropic internal rotation about an axis which also undergoes overall rotatory diffusion. This model can successfully account for the carbon-13 $T_1$ and NOE values observed for the methyl carbons in PIB (5). The methyl group is

pictured as rotating about the three-fold symmetry axis which is
rigidly attached to a backbone bond undergoing isotropic tumbling
caused by backbone rearrangements.  On the other hand, this model
failed to account for relaxation of nuclei in the phenyl group of
polystyrene (6).

It was soon realized that a distribution of exponential correlation times is required to characterize backbone motion for a
successful interpretation of both carbon-13 $T_1$ and NOE values in
many polymers (1, 10).  A correlation function corresponding to a
distribution of exponential correlation times can be generated in
two ways.  First, a convenient mathematical form can serve as the
basis for generating and adjusting a distribution of correlation
times.  Functions used earlier for the analysis of dielectric relaxation such as the Cole-Cole (11) and Fuoss-Kirkwood (12) descriptions can be applied to the interpretation of carbon-13
relaxation.  Probably the most proficient of the mathematical form
models is the log-$\chi^2$ distribution introduced by Schaefer (10).
These models are able to account for carbon-13 $T_1$ and NOE data although some authors have questioned the physical insight provided
by the fitting parameters (11, 13).

The second method used to generate correlation functions
which result in a distribution of exponential correlation times is
to start with a lattice model and consider rearrangements caused
by a crankshaft motion, the three-bond jump.  There are now at
least three modifications of this model all based on the approach
introduced by Valeur, Jarry, Geny and Monnerie (VJGM)(14, 15, 16).
The fundamental correlation function is identical with one developed by Glarum (17), and Hunt and Powles (18) from a different
physical picture.  This second type of model involving a distribution of correlation times has been successful in interpreting
carbon-13 relaxation data (11), and it has also been successful in
interpreting proton relaxation data or proton and carbon relaxation data which were not interpreted as well by the distributions
of correlation times generated from mathematical forms (2, 8).  In
addition to these interpretational advantages, the lattice models
may provide a better basis for physical insight since they are
based on a specific motion possible in linear polymers (13) and
since the dependence of model parameters on concentration and temperature seems reasonable (11).  Because of these potential advantages, we shall turn our attention to a more detailed consideration
of the lattice models.

Lattice Models

The starting point of all three models is the equation

$$dP_a^n/dt = w(-2P_a^n + P_a^{n-2} + P_a^{n+2})$$

[2]

which expresses the time dependence of the probability P of bond n in the a direction. The probability per unit time that any particular three-bond segment with the proper gauche conformation undergoes rearrangement is w. For a very long chain, a continuous solution was produced by VJGM (14, 15) which served as a basis for a correlation function and, by Fourier transformation, a spectral density. However, this result alone did not provide a consistent interpretation of spin relaxation until the correlation function was modified by the inclusion of an additional exponential decay (16, 19). The modified spectral density, most simply written as

$$
J(\omega) = \frac{\tau_0 \tau_D (\tau_0 - \tau_D)}{(\tau_0 - \tau_D)^2 + \omega^2 \tau_0^2 \tau_D^2} \left\{ \left( \frac{\tau_0}{2\tau_D} \right)^{1/2} \right.
$$

$$
\times \left[ \frac{(1 + \omega^2 \tau_0^2)^{1/2} + 1}{1 + \omega^2 \tau_0^2} \right]^{1/2} + \left( \frac{\tau_0}{2\tau_D} \right)^{1/2} \qquad [3]
$$

$$
\times \frac{\omega \tau_0 \tau_D}{(\tau_0 - \tau_D)} \left[ \frac{(1 + \omega^2 \tau_0^2)^{1/2} - 1}{1 + \omega^2 \tau_0^2} \right]^{1/2} - 1 \right\}
$$

has two adjustable parameters. The rate of occurrence of the three-bond jump is governed by the choice of $\tau_D$, and the added exponential decay is governed by $\tau_0$. The physical significance of $\tau_0$ is of some interest since it was not introduced from the fundamental lattice equation, Eq. 2.

In interpretational applications it is found that $\tau_D$ and $\tau_0$ have different apparent activation energies (2, 20). The activation energy for $\tau_0$ is lower than for $\tau_D$ and has values nearly equal to the activation energy derived from solvent viscosity (2, 20). The larger activation energy of $\tau_D$ has been associated with backbone rearrangements while that of $\tau_0$ has been associated with long range tumbling. Going back to the lattice equations, it is not easy to support such a distinction, but on the other hand $\tau_0$ does enter the correlation function in the same manner as an entirely independent overall rotatory diffusion correlation time.

A second solution to Eq. 2 is derived from a different physical picture. Jones and Stockmayer (13) solved the lattice equation for a finite segment length. Rearrangements caused by the three-bond jump are considered in a segment containing 2m-1 bonds with complete neglect of directional correlations of bonds outside the segment. This yields a dynamic description for the central bond in the segment which has been carried through to produce the spectral density (13),

$$J(\omega) = 2 \sum_{k=1}^{s} \frac{G_k \tau_k}{1 + \omega^2 \tau_k^2}$$

[4]

$$\tau_k^{-1} = w\lambda_k \qquad s = \frac{m + 1}{2}$$

$$\lambda_k = 4 \sin^2((2k-1)\pi/2(m+1))$$

$$G_k = 1/s + (2/s) \sum_{q=1}^{s=1} \exp(-\gamma q) \cos((2k-1) \pi q/2s)$$

$$\gamma = \ln 9.$$

In this model the parameters are the segment length governed by the choice of 2m - 1, and the rate of occurrence of the three-bond jump governed by the choice of w but usually expressed by $\tau_h$ which equals $(2w)^{-1}$. For the first time, the form of the model given here allows for any choice of m. Applications of this model have yielded sensible values for an apparent activation energy for backbone rearrangements based on $\tau_h$ (6, 8, 21). Over temperature intervals of 50° to 100°C, the segment length remains constant with values of the order of 5 to 15 bonds for polymers in solution, and segment length has been considered as a measure of length of chain involved in cooperative or coupled motions (6, 8, 13, 21). The Jones and Stockmayer model is not a continuous solution and becomes cumbersome for segment lengths of the order of hundreds which is encountered in some but not all solid rubbers.

The third lattice solution presented by Bendler and Yaris (22) allows the number of bonds to become a continuous variable with the parameters of the model being both a short and long range cutoff of motions expressed as an upper and lower frequency, $W_A$ and $W_B$, respectively. The spectral density expression is

$$J(\omega) = (2/(\omega^{1/2}(W_B^{1/2} - W_A^{1/2}))) \int_{(W_A/\omega)^{1/2}}^{(W_B/\omega)^{1/2}} X^2/(1 + X^4)\, dX$$

$$X = k(W_2/\omega)^{1/2}$$

[5]

$$W_2 = 2w/5$$

$w$ = three-bond jump rate.

When this model is applied to the interpretation of spin relaxation of polymers in solution, the extent of cooperation motion can be measured by a parameter $R = (W_A/W_B)^{1/2}$ which is found to take on values from 1 to 50. If a bond is the smallest moving unit, then R, called the "range", corresponds approximately to the number of bonds involved in cooperative or coupled motion (22). Both $W_B$ and $W_A$ are strongly temperature dependent and vary non-monotonically with temperature which appears to complicate this simple identification of R.

Table I.

COMPARISONS OF INTERPRETATIONS OF METHINE CARBON
RELAXATION IN DISSOLVED POLYSTYRENE

| | $T_1$ (ms) | $T_2$ (ms) | NOE | Model Parameters |
|---|---|---|---|---|
| Experiment (10) | 65 | 26 | 1.8 | |
| VJGM (22) | 66 | 25 | 1.8 | $\tau_0 = 3.5 \times 10^{-8}$ s |
| | | | | $\tau_D = 1.9 \times 10^{-9}$ s |
| Jones and Stockmayer | 65 | 28 | 2.07 | $\tau_h = 1.0 \times 10^{-9}$ s |
| | | | | $(2m - 1) = 17$ |
| Bendler and Yaris (22) | 65 | 26 | 2.1 | $W_A = 6.3 \times 10^6$ Hz |
| | | | | $W_B = 4 \times 10^9$ Hz |
| | | | | $R = 16.4$ |

Following Bendler and Yaris (22), it seems fruitful to compare the results of fitting the three lattice models to a common data set.  A standard choice would be the experimental values obtained by Schaefer (10) on isotactic polystyrene in o-dichlorobenzene at 35°C.  Table I contains the fitting parameters for the VJGM and Bendler and Yaris models (22) combined with the Jones and Stockmayer approach.  All three models can account for the data within experimental error, and the only immediate comparison of model parameters is between the range·of the Bendler and Yaris model and the segment length of the Jones and Stockmayer model. Given the rather different mathematical approaches of the two models, the range and segment length are surprisingly similar.

To compare the time scales of the dynamics characterization produced by each model, the spectral density or correlation function can be written as a distribution of exponential correlation times.  For a correlation function, $\Phi(t)$, the general expression is

$$\Phi(t) = \int_0^\infty G(\tau) \exp(-t/\tau) \, d\tau \qquad [6]$$

except in the Jones and Stockmayer model where the integral is replaced by a sum.  With expressions given by the authors of each model, one can calculate the weighted inverse first moment from

$$(\tau_h^*)^{-1} = \int_0^\infty \tau^{-1} G(\tau) \, d\tau \qquad [7]$$

which corresponds to the weighted harmonic average correlation time $\tau_h^*$.  Previously in the Jones and Stockmayer model, the unweighted harmonic average correlation time was labeled $\tau_h$.  For the VJGM model, the inverse first moment is infinite which indicates the presence of very many short correlation times produced by the continuous solution.  For the Jones and Stockmayer model, a summation corresponding to Eq. 7 yields a value of 0.68 ns for $\tau_h^*$ based on the polystyrene interpretation.  In the case of the Bendler and Yaris model, Eq. 7 reduces the simple expression

$$\tau_h^* = 3/(W_A + W_B + (W_B W_A)^{1/2}). \qquad [8]$$

The value of $\tau_h^*$ calculated from $W_B$ and $W_A$ corresponding to the polystyrene interpretation is 0.72 ns which is in good agreement with the time scale of the Jones and Stockmayer model.

One can also compare the spectral densities for the three models as was done by Bendler and Yaris for their model relative to the VJGM model (22).  The comparison of the spectral densities produced by the three models using the parameters given in Table I

is shown in Fig. 1.  All three spectral densities are of rather
similar shape in the region of common NMR measurements which leads
to the conclusion that it may not be possible to distinguish between the three lattice models on the basis of interpretive
ability.  The models may be better distinguished through ease of
utilization and physical significance of model parameters.  The
VJGM model is easiest to apply requiring adjustment of only two
continuous parameters in a concise equation.  The Bendler and
Yaris model involves a numerical integration while the Jones and
Stockmayer model, a summation over discrete values.  As mentioned,
the physical significance of $\tau_D$ and $\tau_0$ in the VJGM model is complicated, and the use of a continuous frequency distribution with
an upper and lower cutoff is rather different from common dynamic
characterizations.  The concept of a "range" from the Bendler and
Yaris model or segment length from the Jones and Stockmayer model
is appealing but more applications are required to test any real
significance.

## Comparison of Polymer Dynamics

Since none of the lattice models is now clearly superior, the
choice for interpretation of spin relaxation in polymers is arbitrary.  Familiarity leads us to select the Jones and Stockmayer
model so we will now consider application of this model to several
well studied polymer systems in order to compare dynamics from
polymer to polymer.  Also the equations required to consider anisotropic internal rotation of substituent groups and overall
molecular tumbling as independent motions in addition to backbone
rearrangements caused by the three-bond jump are available for the
Jones and Stockmayer model (13).
Some of the best studied dissolved polymers are polystyrene
(6, 10, 23-28), polyisobutylene (5, 13, 29, 30), and poly(phenylene oxide)(7, 8).  We will also add to these systems polyethylene
which is an interesting reference point for polymers interpreted
from a tetrahedral lattice viewpoint.  The four systems will be
abbreviated as PS, PIB, M2PPO, and PE in respective order of introduction here.  The first of these systems, PS, has data available as a function of molecular weight, temperature, field
strength, concentration and from three different types of nuclei.
Similar large data bases are available for PIB and M2PPO, but the
PE results are quoted from a proton and carbon-13 study as a function of molecular weight and temperature (31).  The PE data are
complicated by an unusual temperature dependence of the relaxation
which precluded any estimation of an apparent activation energy.
Within experimental error, both the proton and carbon-13 spin lattice relaxation times of PE dissolved in p-xylene are independent
of temperature between 90 and 180°C.  The proton and carbon-13
relaxation as a function of molecular weight can be consistently
interpreted in terms of the model employed here, but no account of
the temperature behaviour is offered.

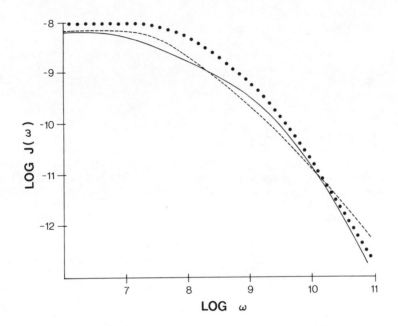

*Figure 1.   Logarithm of the spectral density vs. logarithm of the frequency, ω. The lines are the fit of the spectral density for the various models to relaxation observed in an o-dichlorobenzene solution of isotactic PS: (———), spectral density derived from the Bendler and Yaris model; (– – –), spectral density derived from the VJGM model; and (· · ·), spectral density derived from the Jones and Stockmayer model.*

Table II.

COMPARISON OF BACKBONE DYNAMICS IN DILUTE SOLUTION

| Polymer | Solvent | Temperature (°C) | $\tau_h$ (ns) | $\tau_a$ (ns) | $E_a$ | Segment Length $2m - 1$ |
|---------|---------|------------------|---------------|---------------|-------|-------------------------|
| PE (31) | d-10 p-xylene | 105 | 0.0052 | 0.0104 | - | 5 |
| PIB (27) | CCl$_4$ | 50 | 0.069 | 0.14 | 18 ± 4 | 5 |
| PS (6) | CDCl$_3$ | 50 | 0.24 | 0.72 | 20 ± 7 | 9 |
| M$_2$PPO (8) | CDCl$_3$ | 50 | 1.9 | 3.8 | 25 ± 5 | 5 |

The interpretation of the four systems is summarized in Table II, and the applicability of a lattice model is reasonable since all four backbones can be considered as being at least approximately characterized by a tetrahedral lattice. The correlation time for the three-bond jump, $\tau_h = (2w)^{-1}$, varies considerably among the four systems in general increasing with the size of the three-bond unit. The simple average correlation time, $\tau_a$, also follows the same general trend. Apparent activation energies do not vary so greatly with the size and complexity of the three-bond unit. The segment length for cooperative or coupled motion is relatively short for these simple backbones in dilute solution. Note that in a different solvent, a somewhat longer segment length of 17 bonds produces the best fit for polystyrene (Table I). From another perspective, short segment lengths correspond to rather narrow distributions of correlation times.

Table III.

COMPARISON OF SUBSTITUENT GROUP ROTATION IN DILUTE SOLUTION

| Polymer | Group | T ($^{o}$C) | $\tau_{ir}$ (ns) | $E_a$ (kJ) |
|---------|-------|-------------|------------------|------------|
| PIB (21) | methyl | 50 | 0.21 | $18 \pm 5$ |
| PS (28) | phenyl | 50 | 1 | $20 \pm 5$ |
| $M_2$PPO (8) | phenyl | 50 | 0.23 | $5 \pm 2$ |

In Table III, the characterization of anisotropic rotation of substituent groups is compared for this same collection of polymers. In dilute solution, all three polymers have similar time scales for this motion as indicated by the value of the internal rotation correlation time, $\tau_{ir}$. However, the activation energy for substituent group rotation in PIB and PS is much higher than the activation energy for phenyl group rotation in $M_2$PPO. This indicates some real differences in the nature of the two local motions. The correlation time for phenyl group rotation in polystyrene is rather uncertain since it is calculated from a very small difference between $T_1$ values. This same problem precludes a very accurate estimate of the activation energy although some attempts have been made (24, 27, 28).

The relationship between internal rotation of substituents and backbone rearrangements can be considered from the interpretation. The time scales of anisotropic internal rotation and backbone rearrangements are well separated in $M_2$PPO. In addition, the concentration and temperature dependences of these two quantities are quite different leading us to conclude that the motions are independent. In PIB and PS, internal rotation and backbone

Table IV.

COMPARISON BETWEEN BACKBONE DYNAMICS AND SUBSTITUENT GROUP ROTATION

| Polymer | Nuclei | Dominant Motional Source of Spin Relaxation | Relationship between Backbone Rearrangement and Internal Rotation |
|---|---|---|---|
| PE (27) | Methylene carbon | backbone rearrangement | |
| PIB (21) | Methylene carbon | backbone rearrangement | independent |
| | Methyl carbon | backbone rearrangement | |
| PS (6) | Methylene carbon | backbone rearrangement | ? |
| | Phenyl carbons | backbone rearrangement | |
| M₂PPO (8) | Phenyl carbon | phenyl group rotation | independent |
| | Methyl protons | phenyl group rotation | |

rearrangements are close in time scale, and the activation energy for methyl group rotation in PIB is equal to the activation energy for backbone rearrangement. However, analysis of the concentration dependence of motions in PIB points to the independence of the two motions (21). The time scale of backbone rearrangements changes by orders of magnitude in traversing the concentration range from dilute solution to the bulk while the time scale of methyl group rotation remains almost constant. This interpretation is at odds with one presented by Heatley (30) but data at two field strengths support the interpretation voiced here. Table IV attempts to summarize these comparisons between backbone rearrangements and substituent group rotation. It also lists the motion which is the major source of spin relaxation through modulation of the dipole-dipole interaction. In only M₂PPO does internal rotation of a substituent group, the phenyl group, become the dominant source of spin relaxation.

Table V.

COMPARISON OF BACKBONE REARRANGEMENTS IN SOLIDS

| Polymer | Temperature | $\tau_h$ (ns) | $\tau_a$ (ns) | Segment Length 2m - 1 |
|---|---|---|---|---|
| PE (amorphous) (32) | 45° | .030 | .060 | 5 |
| PIB (rubber) (5) | 45° | 11 | 22 | 5 |
| Cis-Polyisoprene (rubber) (10) | 35° | 0.35 | 5.25 | 57 |
| Cis-Polybutadiene (rubber) (10) | 35° | 0.0065 | 3.7 | $23 \times 10^2$ |

Only recently have the lattice models been applied to solid amorphous and rubbery polymers (21-22). Table V contains a summary of the interpretations of several solid polymers. In general, less extensive data are available on these systems. For PE (32) and PIB (5), the interpretation is based on carbon-13 $T_1$ and NOE values.

For cis-polyisoprene and cis polybutadiene, the interpretation is based on $T_1$, NOE and $T_2$ values (10) although utilization of $T_2$ is complicated because of the possible presence of systematic errors particularly in spectrometers employing superconducting magnets. The interpretation for PE and to a lesser extent PIB is not unique. Several choices of segment length are possible and the parameters listed in Table V are for a minimum length which

accounts for $T_1$ and the NOE.  The other possible choices of segment length are not much longer since a maximum NOE is observed in
PE, and the PIB data are consistent with a very narrow distribution of correlation times (5, 21).  Of course coupling between
chains is completely neglected in this lattice model so the significance of segment lengths involved in cooperative local motions
is unclear.  For cis-polyisoprene and cis-polybutadiene, the
choice of parameters accounting for $T_1$, the NOE, and $T_2$ is unique.
However, the segment length is determined entirely by matching $T_2$
which is the least reliable piece of input information although
great care was taken by Schaefer (10) to limit systematic errors.
In PE, the NOE and apparent $T_2$ depend on the degree of crystallinity (32) and mostly amorphous PE is considered here.  As
crystallinity increases, the NOE and $T_2$ decrease which corresponds
to an increase in coupling segment length.  However, the decrease
in $T_2$ may well reflect macroscopic and microscopic inhomogeneities
in magnetic field strength across the sample caused by instrumen-
tal problems and sample preparation.  These kinds of complications
led us and others (2) to doubt the significance of dynamic interpretations which rest strongly on the observed values of $T_2$.  It is
preferable to develop another measure sensitive to low frequency
motions which is less susceptible to systematic errors.  Possibly
$T_{1\rho}$ could play this role in viscous solution and bulk materials as
it is now providing insight into glassy materials (33).

One last comment on the interpretation of cis-polyisoprene
and cis-polybutadiene is appropriate.  Models based on a tetrahedral lattice of equivalent bonds are clearly not strictly applicable to a polymer containing both single and double bonds.  Thus,
model parameters including the coupling segment length should not
be taken too seriously although it is very nearly equal to the
ranges obtained by Bendler and Yaris employing their model.  One
can be somewhat hopeful about the application of the lattice models to non-tetrahedral backbones if the connectivity of the chain
is the major factor determining the correlation function and not
the tetrahedral geometry (34).

Summary

The lattice models provide useful interpretations of spin relaxation in dissolved polymers and rubbery or amorphous bulk polymers.  Very large data bases are required to distinguish the
interpretive ability of lattice models from other models, but as
yet no important distinction between the lattice models is apparent.  In solution, the spectral density at several frequencies
can be determined by observing both carbon-13 and proton relaxation processes.  However, all the frequencies are rather high
unless $T_2$ data are also included which then involves the prospect
of systematic errors.  It should be mentioned that only effective
rotational motions of either very local or very long range nature
are required to account for solution observations.  The local

motions which result in rotational averaging of dipolar interactions are the backbone rearrangements caused by the three-bond jump or anisotropic rotation of substituent groups. The only long range motion included is overall rotatory diffusion. Together these account for the molecular weight dependence of spin relaxation although mid-range motions corresponding to Rouse-Zimm modes of an order higher than one are not included. The effects of local motions are seen in the data at high molecular weight and the effects of overall rotatory diffusion, at low molecular weights. However, it has not been necessary to explicitly include other types of cooperative, long range motions. Possibly $T_{1\rho}$ would be sensitive to motions corresponding to the higher Rouse-Zimm modes, and this would aid in the elucidation of the role of these motions in spin relaxation.

The other class of motion only now being introduced into interpretive models is oscillatory motion. Anisotropic oscillatory motions of substituent groups have been considered by Chachaty (12) but not in conjunction with a lattice description of backbone motion. No attempt to develop a model based on oscillatory backbone rearrangements is known to these authors, and this avenue may be very important for the interpretation of concentrated solutions, rubbery or amorphous solids, and especially glassy polymers (32).

## Acknowledgments

Many helpful discussions with W. H. Stockmayer are greatly appreciated. The research was carried out with financial support of the National Science Foundation, Grant DMR7716088, Polymers Program. This research was supported in part by a National Science Foundation Equipment Grant No. CHE77-09059.

## Abstract

A brief survey of models for the interpretation of spin relaxation in polymers suggests models based on the occurrence of the three-bond jump on a tetrahedral lattice are capable both of accounting for observations and providing some physical insight. The lattice model of Valeur, Jarry, Geny and Monnerie is compared with the more recent revisions of Jones and Stockmayer, and Bendler and Yaris. Since it is found that all three lattice models have comparable interpretive ability and produce very similar descriptions of the spectral density when applied to the same data, one model, the Jones and Stockmayer version, was used to interpret several well studied polymers. The resulting characterization of motions in dissolved polyethylene, polyisobutylene, polystyrene and poly(phenylene oxide) are reviewed for trends in time scale, apparent activation energy and the extent of cooperative motion. Time scales varied from picoseconds to nanoseconds, the activation energies for backbone rearrangements are all about

20kJ, and the length of chain involved in cooperative motion is only 5 to 15 bonds. Spin relaxation in four solid polymers was also interpreted with the model. Amorphous polyethylene and polyisobutylene rubber undergo motion nearly as rapid as dissolved polymers and the segment length for cooperative motion is not appreciably longer either. Cis-polyisoprene and cis-polybutadiene are also very mobile as solid rubbers but the apparent segment length for cooperative motion is much longer than for simple dissolved polymers. Of course, a tetrahedral lattice model is not strictly applicable to these last two polymers, and interchain cooperativity was not properly considered for any of the solid polymers.

Literature Cited

1.  For a recent overview, Schaefer, J. in "Topics in Carbon-13 NMR Spectroscopy," Vol. 1, G. C. Levy, Ed. (Wiley-Interscience, New York, 1974).
2.  Heatley, F. and Cox, M. K., Polymer, (1977), 18, 225.
3.  Doddrell, D., Glushko, V. and Allerhand, A., J. Chem. Phys., (1972), 56, 3683.
4.  Bloembergen, N., Purcell, E. M. and Pound, R. V., Phys. Rev., (1948), 73, 679.
5.  Komoroski, R. A. and Mandelkern, L., J. Polym. Sci., (1976), Sym. No. 54, 201.
6.  Matsuo, K., Kuhlmann, K. F., Yang, H. W.-H., Geny, F., Stockmayer, W. H. and Jones, A. A., J. Polym. Sci., Polym. Phys. Ed., (1977), 15, 1347.
7.  Laupretre, F. and Monnerie, L., Eur. Polym. J., (1974), 10, 21.
8.  Jones, A. A. and Lubianez, R. P., Macromolecules, (1978), 11, 126.
9.  Woessner, D. E., J. Chem. Phys., (1962), 36, 1.
10. Schaefer, J., Macromolecules, (1973), 6, 882.
11. Heatley, F. and Begum, A., Polymer, (1976), 17, 399.
12. Ghesquiere, D., Ban, B. and Chachaty, C., Macromolecules, (1977), 10, 743.
13. Jones, A. A. and Stockmayer, W. H., J. Polym. Sci., Polym. Phys. Ed., (1977), 15, 847.
14. Valeur, B., Jarry, J. P., Geny, F. and Monnerie, L., J. Polym. Sci., Polym. Phys. Ed., (1975), 13, 667.
15. Valeur, B., Monnerie, L. and Jarry, J. P., J. Polym. Sci., Polym. Phys. Ed., (1975), 13, 675.
16. Valeur, B., Jarry, J. P., Geny, F. and Monnerie, L., J. Polym. Sci., Polym. Phys. Ed., (1975), 13, 2251.
17. Glarum, S. H., J. Chem. Phys., (1960), 33, 639.
18. Hunt, B. I. and Powles, J. G., Proc. Phys. Soc., (1966), 88, 513.
19. DuBois-Violette, E., Geny, F., Monnerie, L. and Parodi, O., J. Chem. Phys., (1969), 66, 1865.
20. Schilling, F. C., Cais, R. E. and Bovey, F. A., Macromole-

cules, (1978), 11, 325.
21. Jones, A. A., Lubianez, R. P., Hanson, M. A. and Shostak, S. L., to appear J. Polym. Sci., Polym. Phys. Ed., (1978).
22. Bendler, J. and Yaris, R., to appear Macromolecules, (1978).
23. Allerhand, A. and Hailstone, R. K., J. Chem. Phys., (1972), 56, 3718.
24. Laupretre, F., Noel, C. and Monnerie, L., J. Polym. Sci., Polym. Phys. Ed., (1977), 15, 2127.
25. Inoue, Y. and Konno, T., Polym. J., (1976), 8, 457.
26. Heatley, F. and Begum, A., Polymer, (1976), 17, 399.
27. Jones, A. A., J. Polym. Sci., Polym. Phys. Ed., (1977), 15, 863.
28. Matsuo, K. and Stockmayer, W. H., private communication.
29. Inoue, Y., Nishioka, A. and Chujo, R., J. Polym. Sci., Polym. Phys. Ed., (1973), 11, 2237.
30. Heatley, F., Polymer, (1975), 16, 493.
31. Gerr, F. E. and Jones, A. A., unpublished results.
32. Komoroski, R. A., Maxfield, J., Sakaguchi, F. and Mandelkern, L., Macromolecules, (1977), 10, 550.
33. Schaefer, J., Stejskal, E. O. and Buchdahl, R., Macromolecules, (1977), 10, 384.
34. Stockmayer, W. H., private communication.

## Discussion

D. Axelson, Florida State University, Florida: One comment and then one question. The polyethylene sample you showed in the last slide I've taken below 45° down to -40°C. The sample develops a broad distribution or non-exponential auto correlation function at low temperatures. One of the giveaways is that in going through the $T_1$ minimum if it is not 110 ms at 67 MHz and if a distribution exists, it will be higher, and in this case the minimum is almost at 300 ms. The system very quickly develops a distribution and the $T_1$'s virtually level off for the last 20 or 35 degrees. Even this macromolecule sample is tremendously complicated. The other point is related to the methyl group rotation you were talking about. Some of the polymers we have been looking at in solution as well as in bulk have long side chains. The end methyl with a six or eight carbon alkyl side chain has a $T_1$ value anywhere from twenty seconds on down. The $T_1$ value is frequency dependent. It shouldn't be frequency dependent over 200 or 300 ms. I was wondering if you have any feel for what physically is taking place or if you have had any opportunity to investigate models that would try to alleviate this peculiarity.

A. Jones, Clark University, Massachusetts: When the methyl group is attached to the backbone or even to a side chain it has a whole distribution of correlation times associated either with the backbone or the side chain in addition to its methyl group rotation. Our assumption in these models is that they are independent so that the methyl group $T_1$ would be calculated on the

basis of a very large distribution of correlation times even
though there probably is a fast correlation time that is dominant.
All those other times are in the correlation function and that may
be what you are detecting.  I'd like to point out that in some
polyethylene data the molecular weight dependence indicated that
at a molecular weight of 150, half of the $T_1$ was still associated
with backbone rearrangements which involves a distribution of
correlation times.  Even a short side chain may have a distribution of correlation times associated with it which you do not
observe until the data is closer to the $T_1$ minimum than in the
polyethylene data considered here.  It is transparent until you
get a sufficient data base.

D. Axelson:  We published a paper (G. C. Levy, D. E. Axelson,
R. Schwartz and J. Hochmann, J. Am. Chem. Soc., 100, 410 (1978))
late last year on methacrylates in which we tried to interpret the
side chains in terms of the effect of the backbone distribution.
The side chain was governed by a combination of the distribution
and the multiple internal rotations.  The problem we had is that
when you get to the end of the chain you still can't account for
the frequency dependence.  We have not proceeded to the stage of
doing dilute solution studies because as you mentioned, there may
be an intermolecular effect with other chains.

A. Jones:  I don't feel the multiple internal rotations model
is applicable for side chains any longer than 3 or 4 carbons.  As
soon as it is longer than 3-4 carbons, a crankshaft type of motion dominates instead of multiple internal rotations.  The viewpoint probably reflects some bias on my part.

J. Prud'homme, University of Montreal, Que.:  What is the
physical explanation for the leveling?  When we see the curve of
$T_1$ as a function of molecular weight there is a plateau.  This
appears at a molecular weight over 1000 and sometimes 10,000.
What is the physical explanation for this?

A. Jones:  With low molecular weights, the overall molecular
tumbling is at a rate not far from the Larmour frequency, the
basic frequency of the experiment.  As molecular weight increases
these overall tumbling motions become very slow and very far removed from the Larmour frequency.  The motion closest to the Larmour frequency comes from backbone rearrangements, very local
motions, which do not depend upon molecular weight.  That three
bond jump in the model does not depend on chain length, it only
involves three bonds in the middle of a long chain.  When those
motions are dominant, the relaxation closest to the Larmour frequency stems from local motions yielding molecular weight independent relaxation.

J. Prud'homme:  What is amazing is that for some substances
this happens at a very high molecular weight.

A. Jones:   If the backbone motion is very slow then higher
molecular weights are required before the overall tumbling motion
becomes much slower than the relatively slow backbone motions.
Dr. Bovey mentioned polysulfones.  I would expect $T_1$ to have an
extended molecular weight dependence because the backbone motions
are relatively slow in this case.

W. G. Miller, University of Minnesota, Minnesota:   In visco
elastic studies the monomeric friction coefficient is used to
describe motion.  The same parameter is used to look at translational diffusion of solvent and its concentration dependence.  Is
there any relationship between this parameter and your three bond
motion or is the correlation length way too long?
A. Jones:   I don't think I can give you a definitive answer.
The three-bond jump should be experiencing a friction partially
determined by its environment, i.e., solvent or other chains
around it, but most probably it is determined by the rotational
potential associated with conformational changes within the chain.
For many systems we see rather little concentration dependency
of the activation energy for these motions.  For instance, the
phenyl group rotation which has a very low barrier appears to be a
good deal more solvent dependent than some of the other backbone
rearrangements involving three bond jumps.  Most others seem to be
mostly influenced by the shape of potential energy surface associated with the chain conformations.

I. C. P. Smith, NRC - Ontario:   When you are testing those
models, crankshafts, cut off and so on, there was a paucity of
data on frequency dependence.  Have you tried to fit the various
models to say three magnetic fields?
A. Jones:   Yes, I didn't show all the data.  Most of the data
is from the literature.  Polystyrene data and polyisobutylene data
at two field strengths can be accounted for by these lattice models based upon the three bond jump.  Some recent data by Dr. Bovey
on the polybutene was more difficult to fit with regard to the
frequency dependence.  (F. C. Schilling, R. E. Cais and F. A.
Bovey, Macromolecules <u>11</u>, 325 (1978).)  Frequency dependent data
I think is important information to acquire when trying to understand the dynamics.

I. C. P. Smith:   But does it distinguish between the models?
It would be nice to find the best of the three.
A. Jones:   It does not distinguish between the lattice models.  I don't think we are going to easily distinguish between the
lattice models.  The same basic equations are employed in each
case.  The three lattice models I mentioned have spectral densities of the same shape.

P. Sipos , Dupont, Ontario:   Using the concept of a chain
segment you have shown large differences between polyisobutylene

and other polymer systems.  Is it that the basic relaxation mode occurs on a long chain because a very short segment oscillates independently or is it a concerted phenomenon?  To get agreement must a large number of segments be considered together?

A. Jones:  We would like to know whether it is a long range oscillation or whether it is a complete turning around of a series of units.  There are measures of whether it is a complete rotation such as the values of $NT_1$ for different carbons.  If $NT_1$ is a con-stant, the motion is probably rotational.  In solution, I think these things tend to be rotational.  As you go to bulk systems and in particular glassy systems, J. Schaefer (J. Schaefer, E. O. Stejskal and R. Buchdahl, Macromolecules 10, 384 (1977) ) feels very strongly that oscillatory motion predominates.  I would agree, although I can't produce an oscillatory model to interpret the data.

RECEIVED March 13, 1979.

# Carbon-13 NMR and Polymer Stereochemical Configuration

JAMES C. RANDALL

Phillips Petroleum Company, Bartlesville, OK 74004

With the discovery of crystalline polypropylene in the early
1950's, polymer stereochemical configuration was established as a
property fundamental to formulating both polymer physical characteristics and mechanical behavior. Although molecular asymmetry
was well understood, polymer asymmetry presented a new type of
problem. Both a description and measurement of polymer asymmetry
were essential for an understanding of the polymer structure.

Technically, each methine carbon in a poly(1-olefin) is asymmetric; however, this asymmetry cannot be observed because two of
the attached groups are essentially equivalent for long chains.
Thus a specific polymer unit configuration can be converted into
its opposite configuration by simple end-to-end rotation and subsequent translation. It is possible, however, to specify relative
configurational differences and Natta introduced the terms isotactic to describe adjacent units with the same configurations and
syndiotactic to describe adjacent units with opposite configurations (1). Although originally used to describe dyad configurations, isotactic now describes a polymer sequence of any number
of like configurations and syndiotactic describes any number of
alternating configurations. Dyad configurations are called meso
if they are alike and racemic if they are unlike (2). Thus from
a configurational standpoint, a poly(1-olefin) can be viewed as a
copolymer of meso and racemic dyads.

The measurement of polymer configuration was difficult and
sometimes speculative until the early 1960's when it was shown
that proton NMR could be used, in several instances, to define
clearly polymer stereochemical configuration. Bovey was able to
identify the configurational structure of poly(methylmethacrylate)
in terms of the configurational triads, mm, mr and rr, in a
classic example (3). In the case of polypropylene, configurational information appeared available but was not unambiguously accessible because severe overlap complicated the identification of
resonances from the mm, mr and rr triads (4). Several papers appeared on the subject of polypropylene tacticity but none totally
resolved the problem (5).

0-8412-0505-1/79/47-103-291$07.00/0
© 1979 American Chemical Society

With the advent of C-13 NMR in the early 1970's, the measurement
of polymer stereochemical configuration became routine and reasonably unambiguous.

The advantages of C-13 NMR in measurements of polymer stereochemical configuration arise primarily from a useful chemical
shift range which is approximately 20 times that of proton NMR.
The structural sensitivity is enhanced through an existence of
well separated resonances for different types of carbon atoms.
Overlap is generally not a limiting problem.  The low natural
abundance (∿1%) of C-13 nuclei is another favorably contributing
factor.  Spin-spin interactions among C-13 nuclei can be safely
neglected and proton interactions can be eliminated entirely
through heteronuclear decoupling.  Thus each resonance in a C-13
NMR spectrum represents the carbon chemical shift of a particular
polymer moiety.  In this respect, C-13 NMR resembles mass spectrometry because each signal represents some fragment of the whole
polymer molecule.  Finally, carbon chemical shifts are well be-
haved from an analytical viewpoint because each can be dissected,
in a strictly additive manner, into contributions from neighboring
carbon atoms and constituents.  This additive behavior led to the
Grant and Paul rules (6), which have been usefully applied in
polymer analyses, for predicting alkane carbon chemical shifts.

The advantages so clearly evident when applying C-13 NMR to
polymer configurational analyses are not devoid of difficulties.
The sensitivity of C-13 NMR to subtle changes in molecular structure creates a wealth of chemical shift-structural information
which must be "sorted out".  Extensive assignments are required
because the chemical shifts relate to sequences from three to
seven units in length.  Model compounds, which are often used in
C-13 analyses, must be very close structurally to the polymer
moiety reproduced.  For this reason, appropriate model compounds
are difficult to obtain.  A model compound found useful in polypropylene configurational assignments was a heptamethylheptadecane
where the relative configurations were known (7).  To be completely accurate, the model compounds should reproduce the conformational as well as the configurational polymer structure.  Thus
reference polymers such as predominantly isotactic and syndiotactic polymers form the best model systems.  Even when available,
only two assignments are obtained from these particular polymers.
Pure reference polymers can be used to generate other assignments
as will be discussed later (5).

To obtain good quantitative C-13 NMR data, one must understand the dynamic characteristics of the polymer under study.
Fourier transform techniques combined with signal averaging are
normally used to obtain C-13 NMR spectra.  Equilibrium conditions
must be established during signal averaging to ensure that the experimental conditions have not led to distorted spectral information.  The nuclear Overhauser effect (NOE), which arises from H-1,
C-13 heteronuclear decoupling during data acquisition, must also
be considered.

Energy transfer, occurring between the H-1 and C-13 nuclear
energy levels during spin decoupling, can lead to enhancements of
the C-13 resonances by factors between 1 and 3. Thus the spectral
relative intensities will only reflect the polymer's moiety concentrations if the NOE's are equal or else taken into consideration.
Experience has shown that polymer NOE's are generally maximal, and
consequently equal, because of a polymer's restricted mobility (8)
(9). To be sure, one should examine the polymer NOE's through
either gated decoupling or paramagnetic quenching and thereby
avoid any misinterpretation of the spectral intensity data.

Let us now discuss C-13 NMR spectra from a series of vinyl
polymers to survey the information available concerning polymer
stereochemical configuration. We will later return to the topics
of how C-13 structural sensitivity is established and how assignments are made. As mentioned earlier, the C-13 configurational
sensitivity falls within a range from triad to pentad for most
vinyl polymers. In noncrystalline polypropylenes, three distinct
regions corresponding to methylene (~46 ppm) methine (~28 ppm)
and methyl (~20 ppm) carbons are observed in the C-13 NMR spectrum. (Throughout this discussion and in the ensuing discussions, the chemical shifts are reported with respect to an internal tetramethylsilane (TMS) standard.) The C-13 spectrum of a
typical amorphous polypropylene is shown in Figure 1.[a] Although
a configurational sensitivity is shown by all three spectral regions, the methyl region exhibits by far the greatest sensitivity
and is consequently of the most value. At least ten resonances,
assigned to the unique pentad sequences, are observed in order,
mmmm, mmmr, rmmr, mmrr, mmrm, rmrr, mrmr, rrrr, rrrm and mrrm,
from low to high field (7) (10) (11). These assignments will be
discussed in more detail later.

The C-13 spectrum of polystyrene, shown in Figures 2 and 3,
contains two regions where stereochemical information can be extracted. There are nine methylene resonances and at least 20–22
aromatic quaternary carbon resonances. No other carbons in polystyrene exhibit a configurational sensitivity. Tentative assignments have been made for the methylene carbons based on an assumed
Bernoullian behavior (12).

As shown in Figure 4, the methylene and methine carbons of
polyvinylchloride show a sensitivity toward configuration and complete, internally consistent assignments have been given by Carman
(13). Likewise a similar situation exists in the C-13 NMR spectrum of polyvinylalcohol, shown in Figure 5, where assignments
have been given by Wu and Ovenall (14). The general trend among
vinyl polymers is for the methylene carbons to exhibit a greater
configurational sensitivity than the methine carbons.

------------------

[a] This spectrum and others shown in this paper were taken from
polymers dissolved in 1,2,4-trichlorobenzene at 125°C.

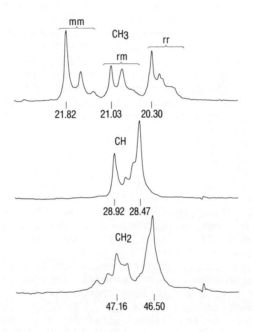

*Figure 1.    Methyl, methine, and methylene regions of the C-13 NMR spectrum of a noncrystalline PP (30)*

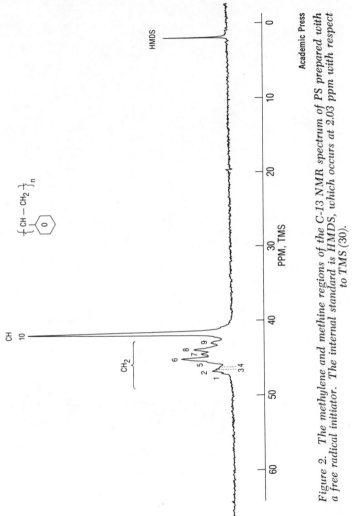

Figure 2.   *The methylene and methine regions of the C-13 NMR spectrum of PS prepared with a free radical initiator.   The internal standard is HMDS, which occurs at 2.03 ppm with respect to TMS (30).*

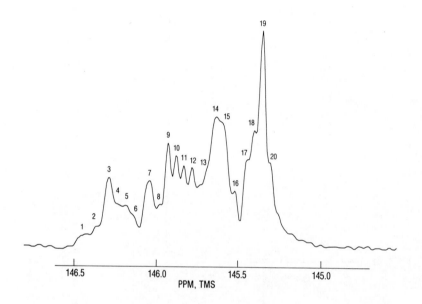

Academic Press

*Figure 3.    The aromatic quaternary resonances from a free radical PS shown in Figure 2 (30)*

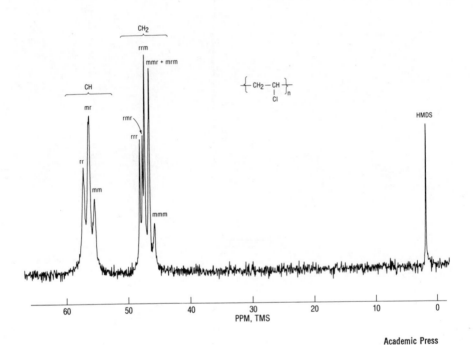

Figure 4.   *C-13 NMR spectrum at 25.2 MHz of PVC (30). See Figure 2 for an explanation of the scale.*

Academic Press

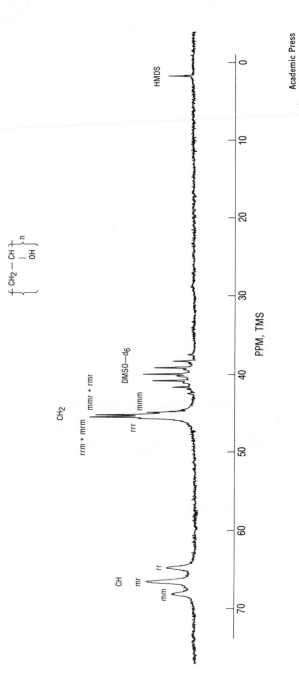

*Figure 5. C-13 NMR spectrum at 25.2 MHz of PVC (30). See Figure 2 for an explanation of the scale.*

Academic Press

Exceptions do exist as shown in the spectrum of polyacrylonitrile in Figure 6. The methine resonances show a distinct triplet with very little splitting exhibited by the methylene resonances. The greatest sensitivity toward configuration occurs for the nitrile resonances where an almost ideal Bernoullian distribution is observed.

In general, the trends observed among configurational assignments cannot be used to make assignments in a system where the assignments are unknown. Certainly, similarities do exist as shown by the data in Tables I, II and III; however, there are no obvious explanations for some of the glaring exceptions. The methine resonances in polyvinylchloride occur as rr, mr and mm from low to high field. A similar trend appears reasonable for polyacrylonitrile; however the methine resonances in polyvinylalcohol have been shown to occur in the inverse order, mm, mr and rr. In an analogous manner, the methine carbon resonances from polypropylene show a pentad sensitivity with the mmmm pentad occurring at low field. In this instance, the observed pentads may occur in an order similar to the methyl pentad resonances (15).

An example where independent, but similar, assignments were obtained occurs for the side-chain carbonyl carbons in polymethylmethacrylate (16) and the quaternary aromatic carbons of poly-$\alpha$-methylstyrene (17). Both are sp$^2$ carbons; otherwise they are in completely different environments. The assignments, as they occur from low to high field, are:

$$\rangle C=O \quad \text{polymethylmethacrylate}$$

mrrm, rrrm, rrrr, rmrm+mmrm, rmrr+mmrr, mmmm, mmmr, rmmr

$$-\underset{|}{C}= \quad \text{poly-}\alpha\text{-methylstyrene}$$

mrrm, rrrm, rrrr, rmrr, mmrm, rmrm+mmrr, mmmm +mmmr +rmmr

Listed in Tables I, II and III are representative examples of the chemical shift behavior for both the backbone methine, methylene carbons and for the various pendant side-chain carbons.

In almost all of the vinyl polymers examined so far, an unambiguous source for configurational information has been found (18). For example, the backbone methylene resonances for the various polyvinylethers show a basic dyad sensitivity with the r resonance occurring downfield from the m resonance (19) (20). In poly-(methylacrylonitrile) one can utilize the methyl resonances (21). The backbone carbons yield configurational information in the C-13 spectra of the various polyalkylacrylates (22) (23). The extraction of configurational information, of course, depends upon the availability of correct assignments in spectra which are often detailed and complex. Let us now turn to the determination of the various spectral sensitivities and the establishment of configurational assignments.

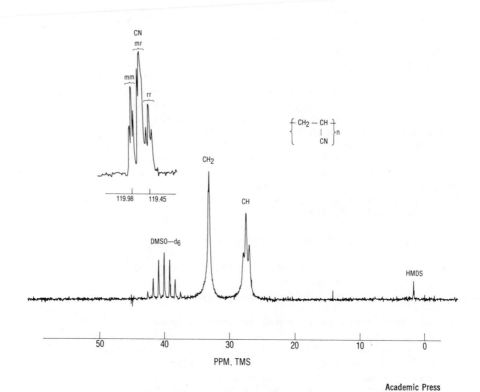

*Figure 6.  C-13 NMR spectrum at 25.2 MHz of PAN (30).  See Figure 2 for an explanation of the scale.*

Table I

The Triad Chemical Shift Sequence with Respect to an Increasing
Field Strength for Some Representative Vinyl Polymer Backbone
Methine Carbon Resonances

$$\delta_{-CH-}$$

| | $\overrightarrow{H}$ | | |
|---|---|---|---|
| Poly(vinylchloride) | rr | mr | mm |
| Poly(isopropyl acrylate) | rr | mr | mm |
| | | | |
| Poly(vinyl alcohol) | mm | mr | rr |
| Polypropylene | mm | mr | rr |
| Poly(vinyl acetate) | mm | (mr) | (rr) |

Polymers with only Singlet Backbone Methine Resonances

    Polystyrene
    Poly(ethyl vinyl ether)
    Poly(isobutyl vinyl ether)
    Poly(methyl acrylate)
    Poly(methyl vinyl ether)

Table II

The Triad Chemical Shift Sequence with Respect to an Increasing
Field Strength for Some Representative Vinyl Polymer Side-Chain
Carbon Resonances

$$\delta_C$$

| | $\overrightarrow{H}$ | | |
|---|---|---|---|
| Polyacrylonitrile    (CN) | mm | mr | rr |
| Poly(tertiary vinyl ether)(-C-) | mm | mr | rr |
| Polypropylene   (CH$_3$) | mm | mr | rr |
| Polystyrene   (-C=) | mm | -- | -- |
| Poly(methyl vinyl ether)   (OCH$_3$) | rr | mr | mm |

## Table III

The Tetrad Chemical Shift Sequence with Respect to an Increasing Field Strength for Some Representative Vinyl Polymer Backbone Methylene Carbon Resonances

$\delta$-CH$_2$- Poly(isopropyl acrylate)

$\overrightarrow{H}$   rrr +   rmr,     mmr +   mrm +   mrr,        mmm

$\delta$-CH$_2$- Poly(methyl acrylate)

$\overrightarrow{H}$   rrr     rmr,     mmr +   mrr,     mrm +   mmm

$\delta$-CH$_2$- Polypropylene

$\overrightarrow{H}$   mrm     rrr     mrr           m

$\delta$-CH$_2$- Polystyrene

$\overrightarrow{H}$   mrr   (rmr)     mmr    (mrm)     mmm     (rrr)

$\delta$-CH$_2$- Poly(vinyl acetate)

$\overrightarrow{H}$   rrr     mrr     mrm +   rmr,     mmr     mmm

$\delta$-CH$_2$- Poly(vinyl alcohol)

$\overrightarrow{H}$   rrr     mrr +   mrm,     rmr +   mmr,     mmm

$\delta$-CH$_2$- Poly(vinyl chloride)

$\overrightarrow{H}$   rrr     rmr     mrr     mrm +   mmr,     mmm

$\delta$-CH$_2$- Poly(ethyl vinyl ether)

$\overrightarrow{H}$       r                         m

$\delta$-CH$_2$- Poly(isobutyl vinyl ether)

$\overrightarrow{H}$       r                         m

$\delta$-CH$_2$- Poly(methyl vinyl ether)

$\overrightarrow{H}$       r                         m

The vinyl polymer studied most thoroughly with respect to
configuration has been polypropylene (5) (7) (10) (11) (24) (25)
(26). The C-13 NMR spectrum of a crystalline polypropylene shown
in Figure 7 contains only three lines which can be identified as
methylene, methine and methyl from low to high field by off-resonance decoupling. An amorphous polypropylene exhibits a C-13
spectrum which contains not only these three lines but additional
resonances in each of the methyl, methine and methylene regions as
shown previously in Figure 1. The crystalline polypropylene must,
therefore, be characterized by a single type of configurational
structure. In this case, the crystalline polypropylene structure
is predominantly isotactic, thus the three lines in Figure 7 must
result from some particular length of meso sequences. This sequence length information is not available from the spectrum of
the crystalline polymer but can be determined from a corresponding
spectrum of the amorphous polymer. To do so, one must examine the
structural symmetry of each carbon atom in the various possible
monomer sequences. Let us begin by an inspection of the polypropylene methyl group in triad and pentad configurational environments. (These arguments can be applied, of course, in a related discussion of any vinyl polymer.) If the methyl group
chemical shift is sensitive to just nearest neighbor configurations, then the simplest configurational sensitivity must be
triad, that is,

There are only three unique triad combinations, mm, mr and rr;
thus a methyl configurational sensitivity to just nearest neighbor
configurations would produce only three resonance in the methyl
region of the C-13 spectrum. From an earlier spectrum of the
amorphous polymer, we noted at least ten methyl resonances. We
must therefore consider the situation where the next-nearest as
well as nearest neighbor configurations are affecting the chemical
shift, that is,

An inspection of the number of unique configurations taken five at
a time (pentads) shows that there are ten unique pentad configurations: mmmm, mmmr, rmmr, mmrm, mmrr, rmr, rmrr, mrrm, mrrr and

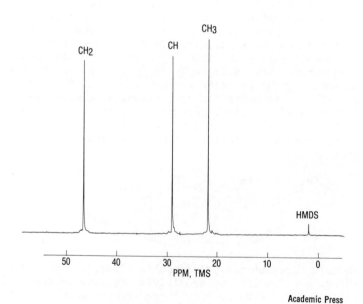

*Figure 7.* C-13 NMR *spectrum at 25.2 MHz of crystalline PP (30). See Figure 2 for an explanation of the scale.*

rrrr. Note that there are three pentads with common mm centers, three with common rr centers and four with common mr triad centers. By now, we can see a pattern developing. The side chain groups of a vinyl polymer will be configurationally sensitive to an odd sequence of structural units; that is, three, five, seven, etc. The number of unique combinations (N) for a particular sequence length (n) can be predicted with the following equation (27):

$$N_{(n)} = 2^{n-2} + 2^{(n-3)/2}$$

Therefore, when

$$n = 3, \qquad N = 3$$

$$n = 5, \qquad N = 10$$

$$n = 7, \qquad N = 36 \qquad etc.$$

These same arguments apply to the polymer backbone methine carbons where the structural sensitivity, required by symmetry, will be to an odd number of continuous units.

In a similar manner, it can be shown that the backbone methylene carbons are sensitive to an even number of structural units with the simplest configurational sensitivity being dyad. The number of unique combinations for a particular sequence length is given by (27):

$$N_{(n)} = 2^{n-2} + 2^{(n-2)/2}$$

where

$$n = 2, \qquad N = 2$$

$$n = 4, \qquad N = 6$$

$$n = 6, \qquad N = 20 \qquad etc.$$

If the structural sensitivity exhibited by C-13 NMR extends beyond next-nearest neighbor configurations, an unwieldy number of configurations must be considered. Such has been the case for the aromatic quaternary carbon resonances in polystyrene where 20-22 resonances are observed as shown in Figure 3. When making these assignments, one must consider thirty-six possible heptads. Even "clearcut" analyses may be deceptively simple. The methyl spectrum of polypropylene apparently consists of three triad regions which are further subdivided into pentads. However, some of the pentads may be composites of overlapping heptads. This is particularly true for the resonances from the rr-centered sequences.

Thus only the assignment problem faced in C-13 NMR polymer configurational studies can be defined by simply counting the number
of observed resonances.  Fortunately, for most vinyl polymers,
the configurational sensitivity is predominantly triad-pentad for
the backbone methine and pendant side-chain carbons and dyad-
tetrad for the backbone methylene carbons.  Caution should be exercised when making assignments because even with the existence
of only a few configurational resonances, the assignments are
often neither straightforward nor unambiguous.

Let us now examine the techniques used in making configurational assignments in polypropylene.  In principle, they could be
applied in any configurational study.  The polypropylene assignments have been well established and are probably the least
equivocal of any reported for vinyl polymers.  Two configurational assignments can be made without difficulty by comparing the
amorphous polymer spectrum with those obtained from reference
polymers consisting predominantly of isotactic and syndiotactic
sequences.  Roberts, et al., identified both the mmmm and rrrr
pentads in the methyl region using this approach (24).  Randall
later correctly identified the configurational sequence of assignments through an extension of the Grant and Paul parameters
to account for configurational differences (10).  Final proof of
the assignments, however, came from a model compound study by
Zambelli, et al., (7) and an epimerization study by Stehling and
Knox (5).  Let's first consider the epimerization study because
it can be more easily adapted to other polymer systems.  An addition of 1% dicumylperoxide plus 4% tris(2,3-dibromopropyl)phos-
phate to a configurationally pure polypropylene results in well-
spaced inversions if the conversions are deliberately kept low.[a]

$$0\ 0\ 0\ 0\ 0\ 0\ 0\ 0\ 0\ 0\ 0 \xrightarrow{\text{epimerization}} 0\ 0\ 0\ 0\ 0\ 1\ 0\ 0\ 0\ 0\ 0$$
$$\text{m m m m m m m m m m} \qquad\qquad \text{m m m m r r m m m m}$$

The new configurational sequences, mmmr, mmrr and mrrm will be
produced from the isotactic polymer by isolated inversions of
configuration.  The relative intensities for the mmmr, mmrr and
mrrm pentads will be 2:2:1 which allows a positive identification
of the mrrm pentad.  Correspondingly, the predominantly syndiotactic polymer will have mm triads inserted into rr.....r sequences by the inversion process.

$$0\ 1\ 0\ 1\ 0\ 1\ 0\ 1\ 0\ 1\ 0 \xrightarrow{\text{epimerization}} 0\ 1\ 0\ 1\ 1\ 1\ 0\ 1\ 0\ 1\ 0$$
$$\text{r r r r r r r r r} \qquad\qquad\quad \text{r r r m m r r r r r}$$

Consequently, the new pentads, mmrr, rrrm and rmmr will be produced in a 2:2:1 ratio.  Once again, the rmmr pentad can be identified by its unique relative intensity.  The mmrr pentad is

-------------

[a] A "0" represents one monomer unit configuration, "1" is its opposite configuration.

commonly produced between the two inversion experiments and therefore can be positively identified. The rrrm and mmmr pentads are
assigned by default. The epimerization experiment, therefore,
leads to an identification of all but the mrmr, mmrm and rmrr pentads. Most importantly, the resonances from common triad centers,
that is, mm, mr and rr have been positively identified. This
triad information is sufficient for a complete determination of
the configurational structure.

The probable "order" of the configurational assignments in
polypropylene have been established by Zambelli, et al., (7) and
A. Provasoli and D. R. Ferro (11) in a study of C-13 labelled compounds, 3(S), 5(R), 7(RS), 9(R͞S), 11(R͞S), 13(R), 15(S)-hepta-
methylheptadecane (compound A) and a mixture of A with 3(S), 5(S),
7(RS), 9(RS), 11(RS), 13(R), 15(S)-heptamethylheptadecane. The
pentad assignments from low to high field are mmmm, mmmr, rmmr,
mmrr, mrrm, rrmr, mrmr, rrrr, mrrr and mrrm in agreement with the
eight assignments by Stehling and Knox and the order predicted
earlier using the NMR additivity relationships. The polypropylene
spectrum appears consistent with the assignment order observed in
the model compounds; however, heptad splitting in the mr and rr-
centered pentads could obscure the true situation and lead to mis-
placed pentad assignments. Inspite of this difficulty, the triad
distribution can be obtained with confidence. Quantitative results can be acquired by either integrating or curve fitting the
methyl region for the mm, mr and rr relative intensities (28).

The most commonly used technique for making C-13 NMR spectral
assignments for vinyl homopolymers has been through the use of
Bernoullian statistics (29) (30). If one knows the relative concentrations of either m or r, any particular "n-ad" distribution
can be calculated because

$$P_m = X_m \tag{3}$$

and

$$1 - P_m = X_r \tag{4}$$

Accordingly, the relative triad and tetrad concentrations are
given by,

| Triad | | Tetrad | |
|---|---|---|---|
| mm | $P_m^2$ | mmmm | $P_m^3$ |
| mr, rm | $2(1-P_m)P_m$ | mmmr, rmm | $2P_m^2(1-P_m)$ |
| rr | $(1-P_m)^2$ | rmr | $P_m(1-P_m)^2$ |
| | | mrm | $P_m^2(1-P_m)$ |
| | | mrr, rrm | $2P_m(1-P_m)^2$ |
| | | rrr | $(1-P_m)^3$ |

If the m and r mole fractions are unknown, one usually iterates
over values for $P_m$ until satisfactory agreement is obtained between the calculated and observed n-ad distributions. The success of the method depends upon the existence of discriminating
differences among the relative intensities and a well-resolved
triad, tetrad or "n-ad" distribution. It is generally best applied when more than one spectral region (for example, a triad
methine distribution and a tetrad methylene distribution) are
available for the calculated versus observed fit. Even under the
best of circumstances, unique fits may not be obtained. If independent partial assignments are available (for example, the isotactic "n-ad" resonance) the method can be applied with more confidence. In cases where Bernoullian fits cannot be obtained, one
then resorts to higher order statistical analyses, that is, first
order Markov or Coleman-Fox (29). Care must be exercised,
however, because more parameters are used to fit the data, thereby
reducing the number of degrees of freedom in the analysis. In
spite of these difficulties, more assignments have been proposed
utilizing conformity to statistical behavior than any other technique. An example where good Bernoullian fits were obtained between the methine and methylene regions has been reported by
Carman for polyvinylchloride (13).

Methine Resonances

| Triad | Observed | Calculated $P_m$ = 0.45 |
|-------|----------|-------------------------|
| rr | 0.291 | 0.303 |
| mr | 0.520 | 0.496 |
| mm | 0.188 | 0.202 |

Methylene Resonances

| Tetrad | Observed | Calculated $P_m$ = 0.45 |
|--------|----------|-------------------------|
| rrr | 0.161 | 0.166 |
| rmr | 0.146 | 0.130 |
| rrm | 0.282 | 0.272 |
| mmr+mrm | 0.320 | 0.334 |
| mmm | 0.092 | 0.091 |

Once assignments are made, C-13 NMR "n-ad" distributions are
available. In general, one would like to obtain a distribution
over the longest possible sequence length. Relationships, often
referred to as the "necessary relationships", exist between n-ad
sequences of different lengths. It is possible to reduce any n-ad
distribution to m versus r, which correspond to the simplest comonomer distribution but is devoid of any information concerning
sequence length.

A few of the necessary relationships, defined by Bovey ($\underline{31}$)$_1$ are

<table>
<tr><td>dyad-triad</td><td>triad-tetrad</td></tr>
<tr><td>m = mm + 1/2 mr</td><td>mm = mmm + 1/2 mmr</td></tr>
<tr><td>r = rr + 1/2 mr</td><td>mr = mrm + 1/2 mrr ⇌ rmr + 1/2 mmr</td></tr>
<tr><td></td><td>rr = rrr + 1/2 mrr</td></tr>
<tr><td></td><td>etc.</td></tr>
</table>

These relationships are also valuable for testing assignments in different regions of the same C-13 NMR polymer spectrum where different configurational sensitivities are shown.

When measuring vinyl polymer tacticity, one prefers the longest complete n-ad distribution available as well as the translated simplest comonomer distribution, possibly m versus r. An alternative exists to the m versus r distribution in the form of number average or mean sequence lengths. If any vinyl homopolymer is viewed conceptually as a copolymer of meso and racemic dyads, mean sequence lengths can be determined for continuous runs of both meso and racemic configurations ($\underline{32}$), that is,

$$\bar{n}_m = \frac{\sum\limits_{i=o}^{i=n} iN_{r(m)_i r}}{\sum\limits_{i=1}^{i=n} N_{r(m)_i r}} = \frac{(mm) + 1/2(mr)}{1/2(mr)} \qquad (5)$$

$$\bar{n}_r = \frac{\sum\limits_{i=o}^{i=n} iN_{m(r)_i m}}{\sum\limits_{i=1}^{i=n} N_{m(r)_i m}} = \frac{(rr) + 1/2(mr)}{1/2(mr)} \qquad (6)$$

In addition to viewing a vinyl homopolymer conceptually as a copolymer of meso and racemic dyads, one may also consider the mean sequence length of "like" configurations ($\underline{28}$). In this instance, the polymer chain is seen as a succession of different lengths of co-oriented configurations from one to "n", the longest sequence of like configurations, that is,

```
r r m m r r r m m m r r m m r m r r r r r m r r r r r m m
1 0 1 1 1 0 1 0 0 0 0 1 0 0 0 1 1 0 1 0 1 0 0 1 0 1 0 1 1 1
  \  /    \  /    \  /  \  /      \  /
 1 1   3  1 1   4   1   3   2 1 1 1 1  2 1 1 1 1   3
```

where,

$$\bar{n}_{like} = \frac{(13)1 + (2)2 + (3)3 + (1)4}{13 + 2 + 3 + 1} = 1.7$$

in the above example.

More generally,

$$\bar{n}_{like} = \frac{\sum_{i=0}^{i=n} iN_{1(0)_i}1 + \sum_{i=0}^{i=n} iN_{0(1)_i}0}{\sum_{i=1}^{i=n} N_{1(0)_i}1 + \sum_{i=1}^{i=n} N_{0(1)_i}0} = \frac{1}{(r)} \tag{7}$$

As was true in the case of mean sequence lengths for meso and racemic dyads, the necessary relationships can be used to develop corresponding equations for any particular n-ad distribution. A favorable point concerning the concept of "like" configurations is that it attaches a physical significance to the racemic distribution.

The mean sequence lengths may offer a better way to present the simple comonomer distribution than meso, racemic distributions because they do reflect the polymer sequential structure. In addition, mean sequence lengths may be useful in evaluating statistical fits. For example,

$$\bar{n}_m = \bar{n}_r = \bar{n}_{like} = 2.0 \tag{8}$$

for ideally random Bernoullian distributions when $P_m = 0.5$. Generally, for Bernoullian distributions,

$$\bar{n}_m = \bar{n}_{like} = 1/(1-P_m) \tag{9}$$

and

$$\bar{n}_r = 1/P_m \tag{10}$$

For first order Markov,

$$\bar{n}_m = 1/(P_{m/r}) \tag{11}$$

$$\bar{n}_r = 1/(P_{r/m}) \tag{12}$$

$$\bar{n}_{like} = 1 + (P_{r/m})/(P_{m/r}) \tag{13}$$

Mean sequence lengths can also be used to write average configurational structures, that is,

$$(m)_{\bar{n}_m} (r)_{\bar{n}_r}$$

In the amorphous polypropylene, shown in Figure 1, the mean sequence lengths are:

$$\bar{n}_{like} = 2.0$$

$$\bar{n}_m = 3.3$$

$$\bar{n}_r = 3.4$$

The sequence distributions are clearly non-Bernoullian in this case although they could satisfy conformity to Markovian behavior as indicated by equations 11-13. The following average structure is suggested by these observed mean sequence lengths.

```
     4    1 1    5     1 1 1    4    1 1
-( 0  0  0  1  0  1  1  1  1  1  0  1  0  1  1  1  1  0  1  0 )_n-
   m  m  m  r  r  r  m  m  m  m  r  r  r  r  m  m  m  r  r  r
```

This amorphous polypropylene, therefore, has a tendency toward short blocks of meso and racemic sequences. This structural conclusion is not readily apparent from a simple inspection of the pentad distribution although the dominant pentads are mmmm, mmmr, rrrr, rrrm and mmrr as indicated by the above formula for the average configurational structure.

In conclusion, we can see that C-13 NMR offers possibly the best experimental approach now available to determine polymer tacticity or stereochemical sequence distributions. The chemical shift sensitivity is generally in the ideal range of dyad to pentad. Higher sensitivities could lead to n-ad distributions which are unwieldy or difficult to assign. From the work reported in the literature, we find that free radical catalysts generally produce configurational distributions which conform to Bernoullian statistics (30). Cationic or anionic catalysts or initiators produce distributions conforming to either Coleman-Fox or higher order Markovian models. Bernoullian statistics have proven to offer a reasonable approach to assignments for polymers produced by free radical initiators. The model compounds for polypropylene have apparently given the correct pentad methyl assignments and may offer a useful approach to assignments in other polymers. Epimerization is one of the more imaginative techniques but so far has seen only a limited application.

When describing polymer tacticity, one should attempt to obtain the highest complete "n-ad" distribution available as well as a simple "comonomer" distribution. In connection with such a measurement, the mean sequence lengths may offer a viable alternative to the simple m versus r distribution. Useful relationships, which are helpful in establishing particular statistical behaviors, are available.

I believe we are just on the threshold of the C-13 NMR applications in studies of polymer chemistry.  Detailed structures are available for correlations with physical properties.  Differences in configurational structure produced by various catalysts can be accurately determined.  For example, the "irregularities" in crystalline polypropylene structure have been shown to be

```
O O O O O O O O O O O O 1 O O O O O O O O O O O
m m m m m m m m m m m r r m m m m m m m m m m m
```

as opposed to

```
O O O O O O O O O O O O 1 1 1 1 1 1 1 1 1 1 1 1
m m m m m m m m m m m r m m m m m m m m m m m m
```

Although not presently reported, the latter structure could be easily detected if it were, in fact, produced by an appropriate catalyst system.

Finally, the configurational structure in copolymers is currently a topic investigated by many industrial scientists.  Is configuration retained when inserting ethylene into a sequence of isotactic propylene units?  This question has been recently answered for a particular catalyst system where it was shown that configuration was retained and catalyst stereospecificity was considered to be the controlling factor (33).  Now that the groundwork has been laid, there should be more extensive and varied applications of C-13 NMR in determining polymer structure.  Certainly, an indispensable tool has become available for determining polymer stereochemical configuration.

## REFERENCES

1.  G. Natta and F. Danusso, J. Polym. Sci., XXXIV, 3 (1959).

2.  F. A. Bovey, "Polymer Conformation and Configuration", Academic Press, New York, N. Y., 1969, p. 8.

3.  F. A. Bovey and G. V. D. Tiers, J. Polymer Sci., 44, 173 (1960).

4.  F. A. Bovey, "Polymer Conformation and Configuration", Academic Press, New York, N. Y., 1969, p. 32.

5.  F. C. Stehling and J. R. Knox, Macromolecules, 8, 595 (1975).

6.  D. M. Grant and E. G. Paul, J. Amer. Chem. Soc., 86, 2984 (1964).

7.  A. Zambelli, P. Locatelli, G. Bajo, and F. A. Bovey, Macromolecules, 8, 687 (1975).

8.  J. Schaefer and D. F. S. Natusch, Macromolecules, 5, 416 (1972).

9.  D. E. Axelson, L. Mandelkern and G. C. Levy, Macromolecules, 10, 557 (1977).

10. J. C. Randall, J. Polym. Sci., Polym. Phys. Ed., 12, 703 (1974).

11. A. Provasoli and D. R. Ferro, Macromolecules, 10, 874 (1977).

12. J. C. Randall, J. Polym. Sci., Polym. Phys. Ed., 13, 889 (1975).

13. C. J. Carman, Macromolecules, 6, 725 (1973).

14. T. K. Wu and D. W. Ovenall, Macromolecules, 6, 582 (1973).

15. J. C. Randall, J. Polym. Sci., Polym. Phys. Ed., 14, 1693 (1976).

16. I. R. Peat and W. F. Reynolds, Tetrahedron Letters, No. 14, 1359 (1972).

17. K. F. Elgert, R. Wicke, B. Stützel and W. Ritter, Polymer, 16, 465 (1975).

18. K. Matsuzaki, H. Ito, T. Kawamura and T. Uryu, J. Polym. Sci., Polym. Chem. Ed., 11, 971 (1973).

19. J. C. Randall, "Polymer Sequence Determination: Carbon-13 NMR Method", Academic Press, New York, N. Y., 1977, Chapter 6.

20. L. F. Johnson, F. Heatley and F. A. Bovey, Macromolecules, 3, 175 (1970).

21. Y. Inoue, K. Koyama, R. Chûjô and A. Nishioka, Makromol. Chem. 175, 277 (1975).

22. H. Girad and P. Monjol, C. R. Hebd. Seances Acad. Sci., Ser. C., 279(13), 553 (1974).

23. H. Matsuzaki, T. Kanai, T. Kawamura, S. Matsumoto and T. Uryu, J. Polym. Sci., Polym. Chem. Ed., 11, 961 (1973).

24.  W. D. Crain, Jr., A. Zambelli, and J. D. Roberts,
     Macromolecules, 3, 330 (1970).

25.  Y. Inoue, A. Nishioka, and R. Chûjô, Makromol. Chem., 152,
     15 (1972).

26.  A. Zambelli, D. E. Dormann, A. I. Richard Brewster and F. A.
     Bovey, Macromolecules, 6, 925 (1973).

27.  F. A. Bovey, "Polymer Conformation and Configuration",
     Academic Press, New York, N. Y., 1969, p. 16.

28.  J. C. Randall, J. Polym. Sci., Polym. Phys. Ed., 14, 2083
     (1976).

29.  F. A. Bovey, "Polymer Conformation and Configuration",
     Academic Press, New York, N. Y., 1969, Chapter 2.

30.  J. C. Randall, "Polymer Sequence Determination:  Carbon-13
     NMR Method", Academic Press, New York, N. Y., 1977, Chapter
     4.

31.  H. L. Frisch, C. L. Mallows, and F. A. Bovey, J. Chem. Phys.
     45, 1565 (1966).

32.  J. C. Randall, "Polymer Sequence Determination:  Carbon-13
     NMR Method", Academic Press, New York, N. Y., 1977, p. 35.

33.  J. M. Sanders and R. A. Komoroski, Macromolecules, 10, 1214
     (1977).

## DISCUSSION

T. K. Wu, Du Pont de Nemours, Delaware:

I have a minor comment concerning the polyvinyl alcohol
spectrum.  The one shown (Figure 5) is similar to one obtained
in our laboratory in 1973.  Last year, Dr. Lana Sheer obtained
a polyvinyl alcohol spectrum where the methine resonance was
resolved into a triplet of triplets.  By studying the effect
of temperature upon the spectrum and by tuning the spectrometer
carefully, one sometimes obtains better hyperfine structure.

J. C. Randall:

Have you obtained satisfactory assignments?

T. K. Wu:

Yes, the assignments have been published in <u>Macromolecules</u> (<u>10</u>, 529 (1977)).

J. C. Randall:

The spectra shown here were obtained from our equipment so, in principle, spectra could be presented which were recorded under similar operating conditions.

W. M. Pasika, Laurentian University, Ontario:

What dictates how far down the chain configurational effects are sensed by the resonance from your reference carbon?

J. C. Randall:

I believe the evidence points to conformational factors. Some people have recently observed chemical shift effects between carbons as far as six bonds away. Originally, Grant and Paul (reference 6) defined chemical shift "contributions" among five neighboring carbons, which were described as $\alpha$ through $\varepsilon$. I believe it is conformational differences which are responsible for the chemical shift behavior observed for various configurations. Depending upon the particular set of circumstances, these effects can involve five to seven consecutive monomer units. Sometimes, these chemical shift differences can be treated by an additive scheme which considers the contributions from each possible configurational array.

W. M. Pasika:

Do the observed differences in configurational chemical shifts disappear at higher temperatures?

J. C. Randall:

There is a temperature effect in many systems. The peaks either collapse or separate. Again, this goes back to the concept that it is the conformation of the configuration that you are dealing with. This is why we see so much unusual chemical shift behavior when we examine trends among vinyl polymers.

G. Babbitt, Allied Chemical, New Jersey:

With regard to branching in polyethylene, I was wondering if you have arranged your experiments in such a way that NOE's and $T_1$'s are taken into account so you can assume

that your areas represent accurate quantitative data.  In
studies of polyethylene branching, there is a strong major
methylene resonance with much weaker methylene and methyl
resonances.  What is the best way to handle such data quantitatively as far as the area measurements are concerned?

J. C. Randall:

     In the systems that I have examined, I can satisfy the
dynamic requirements with a ten second pulse delay.  The longest
methyl $T_1$ may be 3 seconds.  In general, the longer the side
chain, the longer will be the methyl $T_1$.  We will hear more about
this subject later on.  We need not be too concerned about NOE
factors because they are usually full under the experimental
conditions (T = 120-130°C) used for polymer quantitative
measurements.  The $T_1$ problem can be handled, even under non-
equilibrium conditions, by utilizing resonances from the same
types of carbon atoms in a quantitative treatment.  Such an
approach can sometimes lead to more efficient quantitative NMR
measurements.  Adequate pulse spacings will have to be used
whenever one wishes to utilize all of the observed resonances.
Quantitative measurements in branched polyethylenes are very
desirable because this is one of the best applications of
analytical polymer C-13 NMR.

G. Babbitt:

     I was wondering more about the physical problem of
obtaining areas, for example, cutting and weighing, when some
of the methylene resonances may be truncated.

J. C. Randall:

     The problems associated with dynamic range are ones which
we would all like to see solved.  I am using peak heights with
some success.  It may be the best method when the relative
heights are 500:1 or higher.  In any event, one would like to
have an independent measurement or reference standard.  I have
checked with infrared analyses because it can measure the methyl
content.  Unfortunately, IR does not discriminate the polymer
methyl end group from methyls on side-chain short branches.
Nevertheless, with this consideration, I have obtained good
correlations between IR and NMR measurements using NMR peak
heights.  Such a comparison can often serve as a guideline
when measuring an internal branching distribution.  More
confidence is placed in the NMR result, if, overall, similar
results are achieved.  Truncation, which occurs among resonances
with relatively low intensities, is a constant worry.  Another
way of checking for possible truncation in a specific ratio
range is through number average molecular weight measurements on

NBS standard polyethylenes where the $M_n$ is known.   Commercial
polyethylenes may also be useful because the $M_n$ is frequently
in the 10 to 15 thousand range which corresponds to the
sensitivity available with most C-13 NMR spectrometers.

D. E. Axelson, Florida State University, Florida:

We have been investigating branched polyethylenes in
solution and would like to respond to a couple of questions which
were brought up.   In a recent paper in Macromolecules (10, 557
(1977)), we showed that some methyl groups, particularly butyl
and longer, have $T_1$'s as long as 7 seconds; hence 5 $T_1$ could be
of the order of 35 seconds at 120°C in solution.   We have also
measured the NOE's of most of the small branches because of the
high sensitivity of our 270 MHz spectrometer, and, as you
indicated, the NOE enhancements have uniformly been full.   We
have been able to measure peak areas because of our good sensitivity.   Electronic integrations have not given satisfactory
results.   More consistent results have been obtained with a
planimeter.   Dr. Cudby of ICI in England has been very kind to
send us some comparisons between NMR and IR total methyl contents.
Close agreement was obtained when the two experiments were done
carefully.

J. C. Randall:

Have you had an opportunity to evaluate the results from
peak heights versus peak area measurements?

D. E. Axelson:

The peak heights worked out fairly well.   I do not believe
you get into a peak height versus peak area problem if the
resolution is sufficiently high and overlap is not a problem.
We found that results from one spectrum to another were within an
order of one to two branches per thousand carbons as far as
consistency was concerned.   Spectra, which were obtained on other
systems under less than equilibrium conditions, show that the
total methyl content, surprisingly, stays very constant.   You
find that the low end is enhanced and the long branches are
saturated slightly; however, the total remains the same.

J. C. Randall:

Yes, I believe more efficient NMR methods can be developed
by working on systems with known amounts of branching.

D. E. Axelson:

A comparison to IR results can be dangerous in the absence of a good standard for the IR experiment. We ran into a problem where the IR results were one-half of that obtained from NMR. This discrepancy can usually be attributed to the IR method.

J. C. Randall:

Yes, since IR measures total methyl groups, problems can develop if the branching content is comparable to the end group concentration.

RECEIVED March 13, 1979.

INDEX